Lecture Notes in Networks and Systems

Volume 244

The series "Lecture Notes in Networks and Systems" publishes the latest developments in Networks and Systems—quickly, informally and with high quality. Original research reported in proceedings and post-proceedings represents the core of LNNS.

Volumes published in LNNS embrace all aspects and subfields of, as well as new challenges in, Networks and Systems.

The series contains proceedings and edited volumes in systems and networks, spanning the areas of Cyber-Physical Systems, Autonomous Systems, Sensor Networks, Control Systems, Energy Systems, Automotive Systems, Biological Systems, Vehicular Networking and Connected Vehicles, Aerospace Systems, Automation, Manufacturing, Smart Grids, Nonlinear Systems, Power Systems, Robotics, Social Systems, Economic Systems and other. Of particular value to both the contributors and the readership are the short publication timeframe and the world-wide distribution and exposure which enable both a wide and rapid dissemination of research output.

The series covers the theory, applications, and perspectives on the state of the art and future developments relevant to systems and networks, decision making, control, complex processes and related areas, as embedded in the fields of interdisciplinary and applied sciences, engineering, computer science, physics, economics, social, and life sciences, as well as the paradigms and methodologies behind them.

Indexed by SCOPUS, INSPEC, WTI Frankfurt eG, zbMATH, SCImago.

All books published in the series are submitted for consideration in Web of Science.

More information about this series at http://www.springer.com/series/15179

Padmalaya Nayak · Souvik Pal · Sheng-Lung Peng
Editors

IoT and Analytics for Sensor Networks

Proceedings of ICWSNUCA 2021

Springer

Editors
Padmalaya Nayak
Department of Computer Science and Engineering
Gokaraju Rangaraju Institute of Engineering and Technology
Hyderabad, Telangana, India

Souvik Pal
Department of Computer Science and Engineering
Global Institute of Management and Technology
Nadia, West Bengal, India

Sheng-Lung Peng
Department of Creative Technologies and Product Design
National Taipei University of Business
Taipei City, T´ai-pei, Taiwan

ISSN 2367-3370 ISSN 2367-3389 (electronic)
Lecture Notes in Networks and Systems
ISBN 978-981-16-2918-1 ISBN 978-981-16-2919-8 (eBook)
https://doi.org/10.1007/978-981-16-2919-8

This Springer imprint is published by the registered company Springer Nature Singapore Pte Ltd.
The registered company address is: 152 Beach Road, #21-01/04 Gateway East, Singapore 189721, Singapore

Organizing Committee and Key Members

Conference Committee Members

Chief Patrons

G. Ganga Raju, President, GRES, India
G. V. K. Ranga Raju, Vice President, GRES, India

Patrons

M. G. Sekharam, CEO, GRES, India
Jandhyala N. Murthy, Director, GRIET, India
J. Praveen, Principal, GRIET, India

General Chair

D. K. Lobiyal, JNU, Delhi, India

Honorary Chair

Sheng-Lung Peng, National Dong Hwa University, Taiwan
Neel Kanth Grover, IKG Punjab Technical University Kapurthala, India

Convenor

Padmalaya Nayak, GRIET, India

Co-Convenor

K. Madhavi, GRIET, Hyderabad, India
Neeraj Mohan, IKG Punjab Technical University Kapurthala, India
Surbhi Gupta, GRIET, Hyderabad, India

Publication Chair

Sheng-Lung Peng, National Taipei University of Business, Taiwan
Padmalaya Nayak, GRIET, Hyderabad, India
Souvik Pal, Global Institute of Management and Technology, India

Program Chair

Siba K. Udgata, University of Hyderabad, Hyderabad, India

Publicity Chair

G. R. Sakthidharan, GRIET, Hyderabad, India
Sachinandan Mohanty, ICFAI, Hyderabad, India
Tanupriya Choudhury, Univ. of Petrolium and Energy Studies, Dehradun, India

Advisory Committee

Ganapati Panda, IIT, Bhubaneswar, India
K. R. Suresh Nair, Chief Technology Officer, NeST Group
V. Kamakshi Prasad, JNTUH, Hyderabad, India
Atul Negi, University of Hyderabad, Hyderabad, India
S. Krishna Kumar, DRDO, Chennai, India
Siba K. Udgata, Hyderabad Central University, Hyderabad, India
Prakash K. Ray, IIIT, Bhubaneswar, India
S. Viswanadha Raju, JNTUH, Hyderabad, India

TPC Chair

Monika Sachdeva, IKG Punjab Technical University Kapurthala, India

Technical Program Committee

Srinivas Chakravarthy, Kettering University, USA
Sayyad Mohiddin, Think Soft, Australia
Manas Ranjan Patra, Berhampur University, Odisha, India
Elsa Estevez, National University of South Argentina
Ismail Saad, University of Malaysia, Sabah, Malyasia
Wan Abul Rahim Wan Mohd Isa, Univ. Technology, MARA, Malaysia
Raghunandan Reddy Alugubelli, MCC, Univ. of South Florida, USA
Mohammad S Hasan, Staffordshire University, United Kingdom
Md. Ahsan Habib, MBST University, Bangaladesh
Md. Abdur Razzaque, University of Dhaka, South Korea
Selvakumar Samuel, Asia Pacific University and Technology, Malaysia
E. Balamurugan, Bluecrest College, Accra, Ghana
Brijesh Kumbhani, IIT, Ropar, India
Aryabhhatta Sahu, IIT, Guwahati, India
Prasant Kumar Sahu, IIT, Bhubaneswar, India

Chandrashekhar Bhende, IIT, Bhubaneswar, India
T. Ramakrishnudu, NIT,Warangal, India
B. N. Bhandari, JNTUH, Hyderabad, India
Debadatta Pati, NIT, Nagaland, India
Pravati Swain, NIT, Goa, India
Subhransu Ranjan Samantray, IIT, Bhubaneswar, India
Aruna Tiwari, IIT, Indore, India
KameswariChebrolu, IIT, Bombay, India
Rajib. Kumar Panigrahi, IIT, Roorkee, India
DVLN Somayajulu, NIT, Warangal, India
Asit Mohanty, CET, Bhubaneswar, India
Neeraj Mohan, IKGPTU, Jalandhar, India
Diptendu Sinha Roy, NIT, Meghalaya, India
Gayadhar Panda, NIT, Meghalaya, India
Prabeen Kumar Padhy, IIITDM, Jabalpur, India
VivekanandanKaniappan, Bharathiar University, Coimbatore, India
Sraban Kumar Mohanty, IIITDM, Jabalpur, India
K. P. Supreethi, JNTUH, Hyderabad, India
Chapram Sudhakar, NIT, Warangal, India
V. Valli Kumari, Andhra University, India
Karthikeyan S. S., IIITDM, Kanchipuram, India
S. Kanimozhi Suguna, SASTRA DEEMED University, Thanjavur, India
Sudhansu Sekhar Singh, KIIT, Bhubaneswar, India
Niranjan Ray, KITT, Bhubaneswar, India
Ghaida Muttashar Abdulsahib, University of Technology-Iraq, Baghdad, Iraq
M. Nageswar Rao, KL University, India
P. Vidya Sagar, KL University, India
K. Suvarna Vani, VR Siddhartha Engineering College, India
Somya Goyal, SCIT, Manipal University Jaipur, Rajasthan, India

Invited Speaker

Prof. Siba K. Udgata
Professor
School of Computer and Information Sciences
University of Hyderabad (Central University), India

Prof. Sheng-Lung Peng
Professor
Department of Creative Technologies and Product Design
National Taipei University of Business, Taiwan

Dr. Amit K. Mishra
Head
Radar Remote Sensing Group (RRSG)
Electrical Engineering Department
University of Cape Town, South Africa

Dr. S. Krishnakumar
Senior Technical Officer
Defence R&D Organisation (DRDO), Government of India, India

Prof. Manas Ranjan Patra
Professor
Department of Computer Science
Berhampur University, India

Prof. Maninder Singh
Professor and Head
Computer Science and Engineering Department
Thapar Institute of Engineering and Technology, India

Preface and Acknowledgement

The main aim of this proceedings book is to bring together leading academic scientists, researchers, and research scholars to exchange and share their experiences and research results on all aspects of Intelligent ecosystems, data Sciences, and IoT systems. ICWSNUCA 2021 is a multidisciplinary conference organized with the objective of bringing together academic scientists, professors, research scholars, and students working in various fields of engineering and Technology. On ICWSNUCA 2021, you will have the opportunity to meet some of the world's leading researchers, to learn about some innovative research ideas and developments around the world, and to become familiar with emerging trends in Science-Technology. The conference will provide the authors, research scholars, listeners with opportunities for national and international collaboration and networking among universities and institutions for promoting research and developing the technologies globally. This conference aims to promote translation of basic research into institutional and industrial research and convert applied investigation into real-time application.

ICWSNUCA 2021 is hosted by Department of Computer Science and Engineering, Gokaraju Rangaraju Institute of Engineering and Technology, Hyderabad, India. ICWSNUCA 2021 has been held during 26–27th February, 2021, in online mode (ZOOM Platform). The conference brought together researchers from all regions around the world working on a variety of fields and provided a stimulating forum for them to exchange ideas and report on their researches. The proceeding of ICWSNUCA 2021 consists of 45 best-selected papers which were submitted to the conferences, and peer-reviewed by conference committee members and international reviewers. The Presenters have presented through virtual screen. Many distinguished scholars and eminent speakers have joined from different Countries like India, Malaysia, Bangladesh, Sri Lanka, Iraq, Libya, and Taiwan to share their knowledge and experience and to explore better ways of educating our future leaders. This conference became a platform to share the knowledge domain among different countries' research culture. The main and foremost pillar of any academic conference is the authors and the researchers. So, we are thankful to the authors for choosing this conference platform to present their works in this pandemic situation.

We are sincerely thankful to Almighty for supporting and standing at all times with us, whether it's good or tough times and given ways to concede us. Starting from the Call for papers till the finalization of chapters, all the team members given their contributions amicably, which it's a positive sign of significant team works, the editors and conference organizers are sincerely thankful to all the members of Springer especially Mr. Aninda Bose for the providing constructive inputs and allowing an opportunity to finalize this conference proceeding. We are also thankful to Dr. Thomas Ditzinger, Prof. William Achauer, and Prof. Anil Chandy for their support. We are also grateful to Mr. Daniel Joseph Glarance, Project Coordinator, Springer for his cooperation. We are thankful to all the reviewers who hail from different places in and around the globe, who shared their support and stand firm toward quality chapter submission in this pandemic situation.

Finally, we would like to wish you have good success in your presentations and social networking. Your strong supports are critical to the success of this conference. We hope that the participants not only enjoyed the technical program in conference but also found eminent speakers and delegates in the virtual platform. Wishing you a fruitful and enjoyable ICWSNUCA 2021.

Hyderabad, India
Nadia, India
Taipei City, Taiwan

Padmalaya Nayak
Sheng-Lung Peng
Souvik Pal

About This Book

The conference proceeding book is a depository of knowledge enriched with recent research findings. The main focus of this volume is to bring all the computing and communication-related technologies in a single platform. ICWSNUCA 2021 is aimed at providing a platform for knowledge exchange of the most recent scientific and technological advances in the fields of data analytics and computational Science, to strengthen the links in the scientific community. This event aspires to bring together leading academic scientists, researchers, industry persons, and research scholars to exchange and contribute to their knowledge, experiences and research outcome on all the phases of Computer Science, Information technology, and data analytics. This book focuses on the applications, use-cases, architectures, deployments, and recent advances of Wireless Sensor Networks as well as Ubiquitous Computing. Different Research Topics may be illustrated in this book, like Wireless Sensor Networks for the Internet of Things, IoT Applications for eHealth, smart cities, etc., architectures for WSNs and IoT, WSNs hardware and new devices, Low-power wireless technologies, Wireless ad hoc Sensor networks, Routing and data transfer in WSNs, Multicast communication in WSNs, Security management in WSNs and in IoT systems, and Power consumption optimization in WSNs. In recent years, we have been hearing about Internet of Things, Industrial IoTs (IIoTs), Industry 4.0, Smart Factories, Smart HealthCare, Smart Logistic and Supply Chains, Smart Mobility, and Smart Energy, etc. These technologies deliver huge economy and financial benefits and will continue the same in near future. But surprisingly, sensor nodes are the backbone of these technologies and these nodes can be worn, carried, embedded in the environment that can provide interesting contextual information. We find sensors in our daily life, starting from Mobile devices, Automobiles, Industrial automation, Robotic systems, Vehicular Tracking and Management Systems, etc. A significant increase in real-world event monitoring capability with Wireless Sensor Networks will lead to a further evolution of ubiquitous computing. Wireless Sensor Networks continue to attract a lot of attention from the academia and industry among researchers, industrials, equipment and chip manufacturers and service providers for promoting large-scale deployments in many applications, such as environmental monitoring, military and medical surveillance, health and wellness applications, intelligent traffic management and user's tracking and identification, including human–computer interaction. It also provides a premier interdisciplinary platform for researchers, practitioners, and educators to present and discuss the most recent innovations, trends, and concerns as well as practical challenges encountered and solutions adopted in the fields of Computer Science. The primary audience for the book incorporates specialists, researchers, graduate understudies, designers, experts, and engineers who are occupied with research and computer science-related issues. The book will be organized in independent chapters to provide readers great readability, adaptability, and flexibility.

Contents

About the Editors

Padmalaya Nayak is working as Professor in department of Computer Science Engineering in Gokaraju Rangaraju Institute of Engineering Technology, Hyderabad under Jawaharlal Technological University, Hyderabad, since 2009. She completed her doctoral degree from National Institute of Technology, Tiruchirappalli, India in 2010. She has 17 years of teaching and research experience in the area of Ad hoc and Sensor Networks. She has published more than 40 research papers in various International Journals and Conferences. She has also contributed two book chapters to her credit. She has visited many countries to present her research paper in many International Conferences. She has received many National/International Level awards for her academic contributions towards the education system. She is the Reviewer of many IEEE/SPRINGER/ELSEVIER Journals and Conferences. She is the member of IEEE, IETE and IEANG professional bodies. She is also a member of the advisory committee for several conference committees.

Souvik Pal is an Associate Professor and Head of the Computer Science and Engineering Department at the Global Institute of Management and Technology, West Bengal, India. Prior to that, he was associated with Brainware University, Kolkata, India; JIS College of Engineering, Nadia; Elitte College of Engineering, Kolkata; and Nalanda Institute of Technology, Bhubaneswar, India. Dr. Pal received his BTech, MTech, and PhD degrees in the field of Computer Science and Engineering. He has more than a decade of academic experience. He is author or co-editor of 12 books from reputed publishers, including Elsevier, Springer, CRC Press, and Wiley, and he holds three patents. He is serving as a series editor for "Advances in Learning Analytics for Intelligent Cloud-IoT Systems", published by Scrivener Publishing (Scopus-indexed) and "Internet of Things: Data-Centric Intelligent Computing, Informatics, and Communication", published CRC Press, Taylor & Francis Group, USA. Dr. Pal has published a number of research papers in Scopus / SCI-indexed international journals and conferences. He is the organizing chair of RICE 2019, Vietnam; RICE 2020 Vietnam; ICICIT 2019, Tunisia. He has been invited as a keynote speaker at ICICCT 2019, Turkey, and ICTIDS 2019,2021, Malaysia, MAICT 2020, Iraq; ICWSNUSA 2021. His professional activities include roles as associate editor and editorial board member for more than 100+ international journals and conferences of high repute and impact. His research area includes cloud computing, big data, internet of things, wireless sensor network, and data analytics. He is a member of many professional organizations, including MIEEE; MCSI; MCSTA/ACM, USA; MIAENG, Hong Kong; MIRED, USA; MACEEE, New Delhi; MIACSIT, Singapore; and MAASCIT, USA.

Sheng-Lung Peng is a Professor and the director (head) of the Department of Creative Technologies and Product Design, National Taipei University of Business, Taiwan. He received the PhD degree in Computer Science from the National Tsing Hua University, Taiwan. He is an honorary Professor of Beijing Information Science and Technology University, China, and a visiting Professor of Ningxia Institute of Science and Technology, China. He is also an adjunct Professor of Mandsaur University, India. Dr. Peng has edited several special issues at journals, such as Soft Computing, Journal of Internet Technology, Journal of Real-Time Image Processing, International Journal of Knowledge and System Science, MDPI Algorithms, and so on. His research interests are in designing and analyzing algorithms for Bioinformatics, Combinatorics, Data Mining, and Networks areas in which he has published over 100 research papers.

An Interactive Smart Mirror Using Internet of Things and Machine Learning

Keval B. Prajapati, Chintan Bhatt, and Hakima Chaouchi

Abstract With the large-scale improvements in communication technology and easy accessibility of the same has led to the advent of the IoT technology, which in turn has resulted in more, and more devices being IoT enabled. This research paper embarks a next generation smart mirror with the aim of connecting the conventional mirror into the IoT network along with a certain set of features, which were never introduced before. The whole range of features focuses on increasing the utility of daily mirrors. This mirror along with acting as a piece of glass will also be acting as a huge glass notepad upon which the user can leave any handwritten messages which are to be conveyed to others and can never go unnoticed. Keeping in mind the smudging effect and the fingerprint mark that are left behind when the mirror surface is touched, this new feature allows one to write the message on the mirror without touching the mirror surface itself. Additionally, the mirror will also support some of common features as well namely date, weather forecast, news headlines and daily remainders.

Keywords OpenCV · Object detection · Object tracking · Camera · Neural networks · Python

1 Introduction

For more than a decade, the development of applications for the Internet of Things (IoT) [1–5] has mainly concerned the construction of systems for data generation, transmission and analysis for integration into everyday life objects, while the data visualization part has still remained attached to the world of graphical user interfaces,

K. B. Prajapati (✉) · C. Bhatt
Charotar University of Science and Technology, Changa, Gujarat, India
e-mail: 16ce091@charusat.edu.in

C. Bhatt
e-mail: chintanbhatt.ce@charusat.ac.in

H. Chaouchi
Télécom SudParis, Évry, France

P. Nayak et al. (eds.), *IoT and Analytics for Sensor Networks*, Lecture Notes in Networks and Systems 244, https://doi.org/10.1007/978-981-16-2919-8_1

executed primarily on personal devices such as laptops, tablets and smartphones. Recently this is no longer true, and there are numerous examples of things that can also be used to display information.

In the contemporary world, people are living an extremely busy life with a very less time in the morning rush hours. Waking up early, getting ready for jobs, catching up the earliest suitable commuting vehicles, which in turn leaves a very less time to get updated with day-to-day changing scenarios of the state, country or the entire world before stepping out of the house. Even though one can be updated by accessing their smartphones, tablets or PCs but for tech-savvy people it just becomes another task. The best time management strategy is to find time where it was never before. Seamless integration of technology into daily lives has made this possible where not only communicating devices like smartphones or tablets play a role for the purpose but also household items partake in the same. Another important point is in a hectic life people often forget about trivial tasks relating to family, friends or colleagues. Even though leaving the message on sticky notes is a good idea but most of the time it goes unnoticed and the task remains incomplete.

This is where the idea of using the mirror surface as a huge notepad comes into the picture. The mirror is such a household item, which is looked upon often either purposefully or accidentally. Taking the advantage of this fact, makes it possible to use the conventional mirror surface as a notepad upon which the user can leave handwritten messages, which are to be conveyed to others, members of the house or for the user himself as a to-do note. The note-taking feature supports all the basic features a regular notepad supports: write, erase and clear all.

The mirror also has an integrated music player. Unlike the conventional music players, this one is controlled via some simple human hand gestures. Introduction of a music player into the mirror enhances the utility of the mirror. For example, during certain occasions, a person might find himself/herself to sit in front of a mirror for a prolonged period in order to dress up. The person can use this music player housed in the mirror instead of playing music on their smartphones or other devices. The feature of hand gestures enhances the multi-tasking of the user as well as allows seamless integration of the same in day-to-day lives. Unlike the other hand gesture detection algorithms, which are based on real-time object detection, the one that is featured in this system is based on a neural network using Keras and TensorFlow, which makes prediction of the hand gestures with the help of a dataset of hand gestures images. This model is fairly more accurate and robust when compared to other ones.

The system also targets a short span of time, which people use to spend in the front of the mirror for dressing up themselves or making/combing their hair. This next generation mirror not only acts as a conventional mirror, which allows one to look in the mirror for dressing up or combing their hair but will also be displaying all the vital, crisp and easily comprehensible information like daily remainders, daily weather forecast, time and date, and top five news headlines. All this information will be displayed on the same piece of glass upon which the user is looking hence increasing the utility of the daily mirror. Apart from that, all the content (weather forecast, news headlines and daily reminder) displayed will be updated on a regular interval of time so that always most recent information will be available with the system at disposal.

The given set of features will not only increase the utility of conventional mirrors but also will become an effortless and seamless way users can adopt the incremental update in conventional mirrors.

The reminder of this paper is structured as follows. Section 2 presents the research motivation, explaining why this upgrade was necessary. It also shows about the related work that has been done in this field. The design architecture, working and features are described in detail in Sect. 3. Results are discussed in Sect. 4, while Sect. 5 shows conclusion and future enhancement of the paper.

2 Motivation and Related Work

Another version of magic mirror is built by Josep Cumeras I Khan [1], which can be controlled by voice and gestures. This system has its very own OS knows as MirrorOS or MOS, which is responsible for all the administration. Along with that it also has an dedicated app which can govern what is to be displayed on the mirror. This mirror uses a set of motion detectors for gesture control. Another project is HomeMirror by Hannah Mitt of Adafruit Industries [2], which uses an old tablet instead of LCD/LED monitor. It can be considered as one of the cheapest smart mirrors. The tablet will be running an android-based application designed to show date, time and other information.

The core concept behind the note-taking feature introduced is based on object detection and tracking the same. All of this is achieved with OpenCV and webcam. Extensive research has been done in this field. Rajeev Thaware, Gouri Patil and Saurabh Joshi of eInfochips [3] have in their research work for object detection and tracking for Effective Traffic Management System using OpenCV has described in simple and lucid manner the entire procedure for object detection.

3 Architecture, Working and Feature

The hardware part involves designing of the wooden frame, which will house all the components of the mirror. The 2-way mirror used is reflective enough so that one can use it as a normal mirror and at a same time is transparent enough to allow light to pass through when the light source is in close proximity to the mirror. The 2-way mirror is sandwiched between the wooden frame and LCD/LED screen. The screen is connected to the Raspberry Pi module, which serves as the brain of the mirror. A webcam is also connected to the Raspberry Pi, which will be helping in the note-taking feature. Another webcam is responsible for hand gesture detection. These hand gestures will in turn control the music player. Working of both the webcams is completely independent. Both the webcams are positioned on the top of the wooden

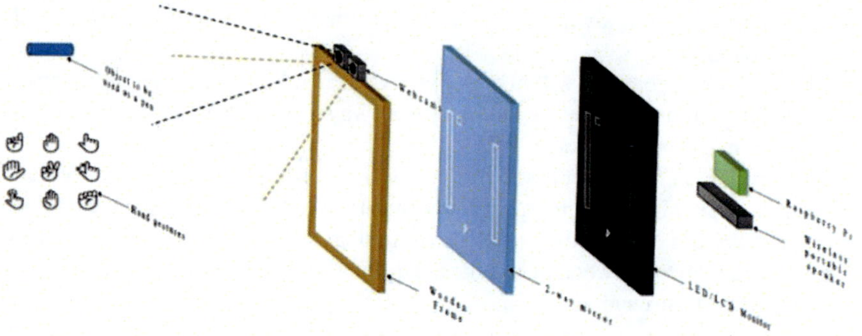

Fig. 1 Hardware model of smart mirror

frame. The system also consists of a wireless speaker, which will be required for music players. Figure 1 shows the hardware design of the next generation mirror.

The software part includes the development of a python application, which will be responsible for displaying the information on to the mirror. The GUI design is mainly done using tkinter library of python. The procedure involved dividing and designing different regions of the screen or mirror for different purposes. The central region of the mirror serves as a notepad. The tkinter library is used to provide software button like write, erase and clear all for this feature. The bottommost region is reserved for news feed, while the top most region is reserved for date, time, weather forecast and remainders. The region between the notepad and news feed is used for displaying the music player. It will consist of three buttons namely six play/pause, next and previous, a marquee text-representing name of current song and a volume slider at bottom of marquee.

3.1 *Hand Gesture-Controlled Music Player*

This is another notable feature of the mirror. Unlike some common hand gesture detection algorithms which are hardly accurate and fail to produce desired results under various scenarios the one which is running for this system is based on a neural network using Keras and TensorFlow which makes prediction of the hand gestures with the help of given dataset of hand gestures images [6]. Initially this model required a statistical training with certain dataset and later on can do complex prediction accurately in different scenarios as well. One can have their own set of gestures for the music control however they need to train the model for the same first. The entire module can be split into 3 parts:

(a) Generate and prepare dataset
(b) Create Model
(c) Predict with Model and give the commands

The following subsections describe each one of them in detail:

(a) Generate and prepare dataset.

The system required a unique set of hand gestures, which can vary up to some extent, so it was required to capture varying images of individual gestures multiple times and create a dataset, which will be in future utilized for training and prediction. The images were captured using a regular webcam, which were later resized, converted from RGB to YCbCr colour and marked pixels of hand with white colour and background with black colour. The dataset consists of 1500 plus images of hand gestures (Fig. 2).

Hand gestures include left, right, palm, fist and victory. Figures 3, 4, 5, 6, 7 depict them, respectively.

(b) Create Model.

It consists of an NN structure with one hidden layer. The size of input layer is nearly 370 k nodes since the size of images is 640 * 576 pixels. The size of the hidden layer is 25 nodes and output will be five nodes (5 types of sign). Figure 8 represents the model. The process includes loading features and labels and randomly initializing NN nodes weights. Then creating cost function, forward propagation and gradient for NN cost function.

(c) Predict with Model and give the commands.

This part involves testing and predicting whether the model is able to recognize the hand gesture in real-time scenarios and able to predict the hand gestures accurately.

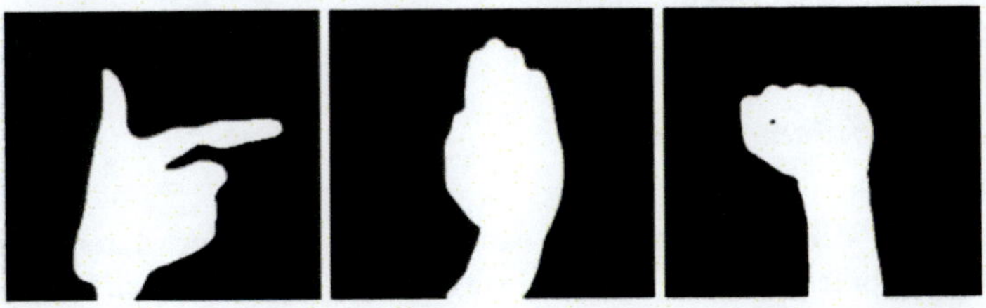

Figs. 2, 3, 4 (2) Right. (3) Palm. (4) Fist

Figs. 5 and 6 (5) Left. (6) Victory

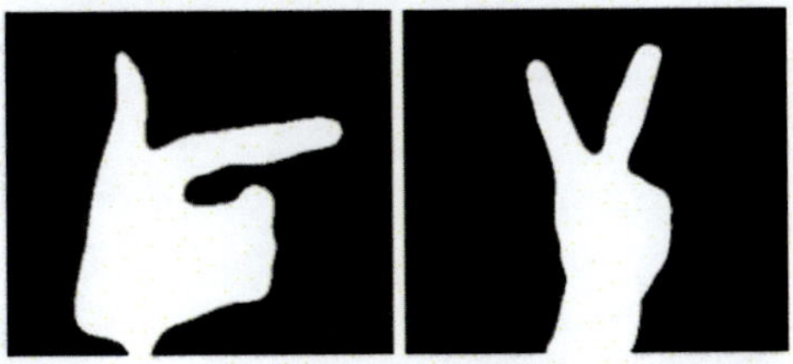

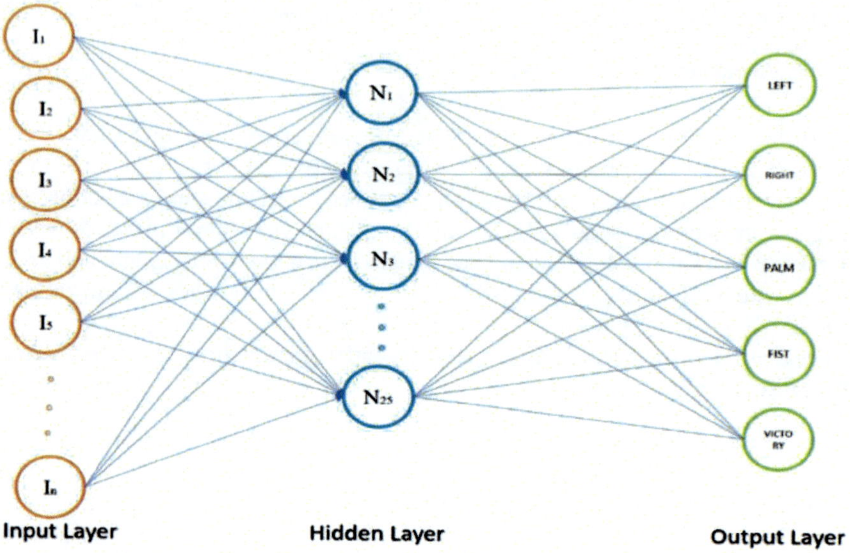

Fig. 7 Neural network model

Once the prediction is done the model should be calling an assigned method in order to control the music player. Here it's

- First to pause the song
- Palm to unpause the song
- Left to play previous song
- Right to play next song
- Victory to increase/decrease the volume.

3.2 *Other Content Features*

Since the mirror will also be displaying date, time, weather forecast, remainders and news feed this content must be delivered via cloud to the application. Date and time are fetched from the Raspbian OS itself while the weather and news are fetched from the cloud. Apart from fetching this content, a request is sent to the cloud on a regular interval of time for updating the content. The remainders are synced with the calendar of the user's smartphone using google API. Figure 8 shows the overall working architecture of the system.

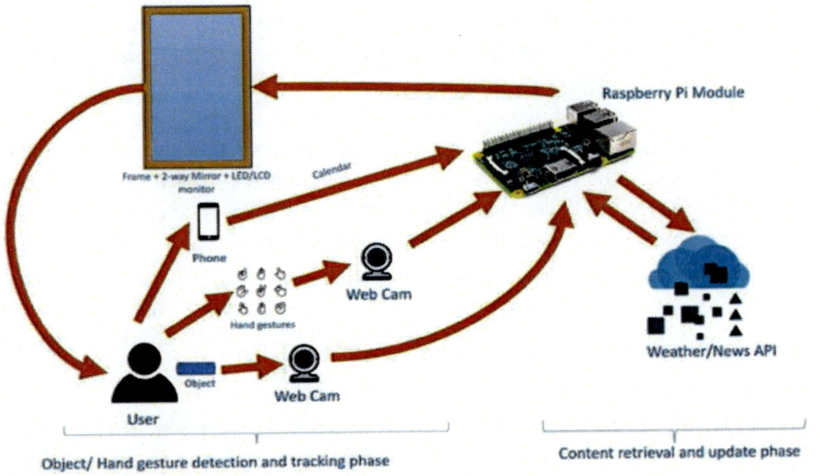

Fig. 8 Working of the next generation mirror

4 Result

With all the components namely Raspberry Pi Module, LCD/LED monitor, two webcams and speakers running all the features of the mirror namely note-taking feature, hand gesture-controlled music player, daily remainders, news feeds, date, time and weather forecast are running perfectly fine. The note-taking features all the components: write, erase and clear all enables its user to use the mirror surface as a huge notepad upon which a handwritten message can be left which is to be conveyed to others. The algorithm for object detection and tracking which are the backbone for note-taking features are working with 100% accuracy. The neural network designed for the hand gesture-controlled music player is working with around 90 to 100% accuracy in predicting real-time hand gestures with help of previously created dataset. Figure 9 shows prediction rate and feature extraction of palm gesture. Once a given hand gesture is predicted an appropriate method is called which in turn governs

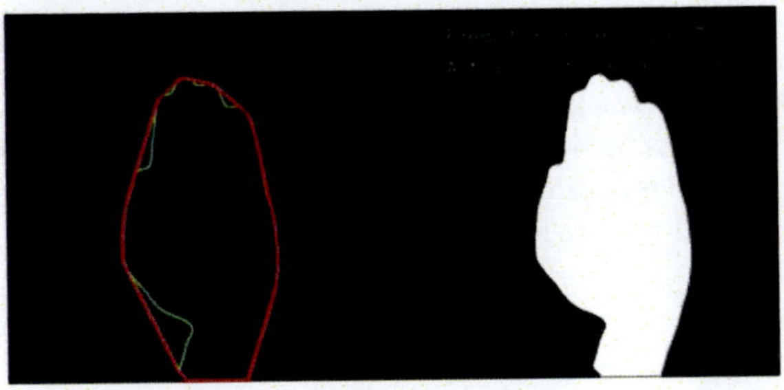

Fig. 9 Palm hand gesture prediction and action

the music player. Apart from this all the basic and common information like daily remainders, date, time, news feed and weather forecast are correct and most recent information. The news feeds are updated every 15 min while the weather forecast is updated every 4 h.

As one can see, the mirror serves as a huge freehand notepad, upon which a message is written. Apart from that, all the three software buttons write, clear all and erase can also be seen. The news feed is present at the bottom while date, time, weather forecast and remainder are present on the top-most region of the mirror.

5 Conclusion and Future Enhancements

This next generation mirror is loaded with features, which are aimed at increasing the utility of the conventional mirror. Apart from that, it also makes it possible to introduce mirrors into the proliferating network of IoT. Since a mirror is inarguably, either the best object to catch attention intentionally or accidentally the mirror will be serving as an excellent notepad upon which the user can leave hand-written messages, which are to be conveyed to others or to the person himself as a to-do work. In order to do so all a person needs to do is to hold a pen and start writing in the air, right in front of the mirror without actually touching the mirror surface itself. The camera placed above the mirror will be tracking the movement of the pen and gives the message as an output. This will also keep the mirror free from smudges and fingerprints. The messages once written could also be erased or be entirely wiped off so that the user can write new messages. Keeping a hand gesture-controlled music player not only enhances multi-tasking ability of the user but it also seamlessly integrates the music player into the mirror hence increasing its utility as well. In addition to this, the features of the mirror also focus on effectively utilizing the short span of time people spend in front of the mirror during morning rush hours by displaying vital information in a very precise and crisp format. This information includes date, time, current weather, news headlines and calendar reminders. The user will be updated with all this information in a short span of time.

The future enhancements include introduction of another feature to the mirror, which is voice control. This voice control feature will be governing a certain new set of features such as maps that can be displayed on the mirror, smartphone notifications or playing a video itself. The proposed system either can be achieved with help of third party software like Google Assistant or Amazon Alexa or can develop a less intelligent voice recognition system, which can follow only a certain set of commands. Apart from that, some iterative upgradation can include improvement in prediction rate of hand gesture recognition system.

References

1. Cumeras I Khan, J. (2016). I built a smart mirror with voice and gesture control and a full OS to go with it. https://imgur.com/gallery/j0BFU.
2. Adafruit. https://learn.adafruit.com/android-smart-home-mirror/overview.
3. Thaware R., Patil, G., & Joshi, S. (eInfochips). Designing an effective traffic management system through vehicle classification and counting techniques. https://www.design-reuse.com/articles/45224/traffic-management-system-through-vehicle-classification-and-counting-techniques.html.
4. Dey, N., Hassanien, A. E., Bhatt, C., et al. (2018). *Internet of things and big data analytics toward next-generation intelligence*. Cham: Springer International Publishing.
5. Bhayani, M., Patel, M., & Bhatt, C. (2016). Internet of things (IoT): In a way of smart world. In: S. Satapathy, Y. Bhatt, A. Joshi, D. Mishra (Eds.), *Proceedings of the International Congress on Information and Communication Technology. Advances in Intelligent Systems and Computing* (Vol. 438). Singapore: Springer. https://doi.org/10.1007/978-981-10-0767-5_37.
6. Kumari, S., Mathesul, S., Shrivastav, P., & Rambhad, A. (2020). Hand gesture-based recognition for interactive human computer using tenser-flow. *International Journal of Advanced Science and Technology, 29*(7), 14186–14197.

Cluster Formation Algorithm in WSNs to Optimize the Energy Consumption Using Self-Organizing Map

Padmalaya Nayak, GK. Swetha, Priyanka Kaushal, and D. G. Padhan

Abstract Wireless Sensor Networks (WSNs) are considered as one of the most prevailing technologies in today's world due to the diversified applications. These applications are huge in the range such as environmental monitoring, health care, civil and military, disaster management to other surveillance systems. Minimization of energy is one of the most exciting tasks in WSNs as small sensor nodes are battery powered and deploy in remote environments. Clustering is one such imperative technique that can conserve energy more broadly, and evenly which makes the network operational for a longer period. This research paper aims at developing an energy-aware cluster-based routing protocol using Artificial Neural Network (ANN) which finds an optimal number of clusters and rotates the cluster head periodically to balance the energy consumption throughout the network. Typically, the proposed algorithm is developed based on Self-Organizing Map (SOM) to form the clusters, and the K-means algorithm is used to form different sizes of clusters. Finally, an optimal number of clusters is found that impacts the network load and balances the energy consumption. MATLAB is used as a simulation tool for experimental analysis. Simulation results prove that the proposed cluster-based routing protocol "LEACH-SOM" dominates the traditional LEACH routing protocol in terms of minimal energy consumption and makes the network active for long period.

Keywords WSN · SOM · K-means · LEACH

P. Nayak · GK. Swetha (✉) · D. G. Padhan
Gokaraju Rangaraju Institute of Engineering and Technology, Hyderabad, India

P. Nayak
e-mail: padmalaya@griet.ac.in

P. Kaushal
Chandigarh Engineering College, Landran, Mohali, India

P. Nayak et al. (eds.), *IoT and Analytics for Sensor Networks*, Lecture Notes in Networks and Systems 244, https://doi.org/10.1007/978-981-16-2919-8_2

1 Introduction

Wireless Sensor Networks (WSNs) are considered as an active emerge area because of vital role in numerous applications. These sensor nodes are battery operated and these batteries are required to recharge and replace very often. So, minimal energy consumption of sensor nodes is the most challenging task. Many existing routing protocols discussed in the literature are energy efficient [1–5] rather than simply routing data from source to destination. Majorly of the routing protocols are classified into flat-based, hierarchical-based, query-based, location-based, etc. The hierarchical routing protocols are the most energy-efficient routing protocols [1–5] because they accommodate large-scale sensor nodes by reducing redundant data transmissions and thus, reduce the energy consumption. As our focus on improving the performance of hierarchical routing protocols, we have considered LEACH protocol as the reference protocol and incorporated machine learning techniques in it. Machine Learning (ML) is the subcategory of Artificial Intelligence (AI), which can learn by itself and improve the system performance without being programmed explicitly. ML techniques have been used for many applications in WSNs. Few applications of ML in WSNs are discussed below [8].

(1) *Scalability*: The routing protocols in WSNs must be scalable because the quantity of sensor nodes varies based on the applications to handle the dynamics of the events. With the help of ML techniques, routing protocols are proved to be scalable.
(2) *Energy Efficiency*: Energy minimization of sensor nodes is an important task in WSNs because small sensor nodes are battery operated. Even though the nodes are energy harvested nodes, it is very important to know when to harvest. The machine learning techniques help to handle the energy conservation issues in WSNs efficiently.
(3) *Faulty tolerance*: The operation of WSN should not be disturbed by faulty or failed sensor nodes. ML techniques help to find the faulty nodes in WSNs.
(4) *Localization*: In WSNs, the sensor nodes are deployed in dynamic environments like underwater, forest, cold places, etc. So, it is very important to identify the location of sensor nodes in any type of dynamic environment. Machine learning techniques help to solve localization issues in WSNs.
(5) *Quality of service*: In many applications, the quality of service is very important that implies the data should be delivered on time without any noise. The machine learning techniques handle efficiently to achieve the quality of service in WSNs.
(6) *Routing*: Routing protocols in WSNs not only handles routing but also handles many problems like energy conservations, faulty nodes, heterogeneity of sensor nodes, etc. So, the development of routing protocols is a very important task in WSNs. Machine learning techniques help to make routing protocols energy efficient and to handle these issues easily.

The remaining section of the paper is partitioned as follows. Section 2 discusses the current literature that contains both the traditional clustered-based routing protocols as well as SOM-based routing protocols in WSNs. Section 3 presents the proposed work followed by the experimental results discussed in Sect. 4. Section 5 concludes the paper.

2 Existing Literature

This section gives an overview of traditional hierarchical routing protocols and few ML-based routing protocols. As our proposed work is based on Self-Organizing Map (SOM), we have discussed a few SOM-based routing protocols in WSNs.

(A) *Hierarchical Routing Protocol.*

The routing protocols in WSNs are categorized in various ways such as based on the network structures, routing operations, and network applications. But majorly of the routing protocols are grouped into flat-based, hierarchical-based, and location-based considering the structure of the network. Hierarchical routing protocols include partition of sensor networks into multiple groups. Each group is controlled by a leader node known as a Cluster Head (CH) as presented in Fig. 1. All the non-cluster nodes that communicate with their CH and CHs are liable to accumulate the sensed data received from the cluster members and communicate with the Base Station (BS). Formation of clusters, selection of CHs, and selection of optimal path are the main challenging tasks in clustered-based routing.

In this regard, LEACH [1] is the oldest hierarchical routing protocol that operates in terms of rounds. Each round carries periodical transmission of messages. It operates in phase manner such as setup phase followed by the transmission phase. Setup phase takes care of the establishment of clusters and cluster head selection. To satisfy this requirement, it uses a probabilistic model. Transmission phase follows

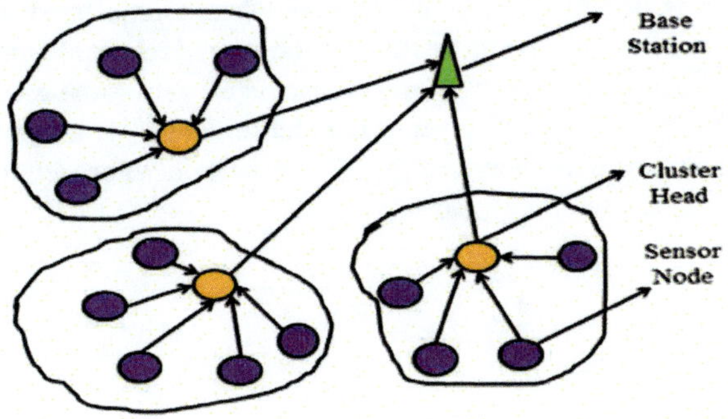

Fig. 1 Basic example of clustered-based WSN

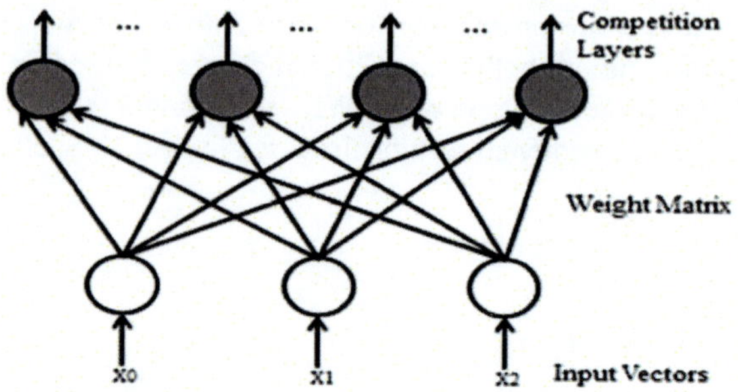

Fig. 2 Architecture of a simple self-organizing map

the setup phase by allowing non-CH nodes to transmit the sensed data to their corresponding CHs. Finally, the CHs take the responsibility to gather the data and deliver it to the Sink node. The authors of [3] have introduced a routing protocol called PEGASIS, where a chain is formed among the sensor nodes through which each sensor node passes the data to the other sensor node till it moves to sink node. TEEN [4] introduces a dynamic routing protocol in which sensor nodes continuously sense the environment and transmits the data till the sensed data exceeds the specified threshold value. APTEEN [5] is the enhanced version of TEEN protocol which is applicable to both reactive and proactive networks. HEED [6] uses a novel clustering approach for routing where the sensor nodes are non-uniformly distributed and has proved its energy efficiency in prolonging the network lifetime compared to other existing protocols.

(B) *SOM-based Routing Protocol.*

Self-Organizing Map (SOM) belongs to unsupervised machine learning technique that allows grouping the unlabeled data according to their similarities, patterns, and differences without any prior training. Unsupervised learning is also known as competitive learning or self-learning developed by Kohonen in 1990. The simple architecture of Self-Organizing Map neural network is shown in Fig. 2. The SOM consists of neurons that are structured in a low-dimensional grid, in which each neuron is connected to other neurons with a neighborhood relation. The SOM consists of the input layer, each input layer neuron is associated with some weight and activities the other neuron in the output layer or competition layer with similarity function called Euclidian distance as shown in Eq. 1. The neuron whose Euclidian distance is minimum becomes a winning neuron.

$$ED_j = \sum_{i=1}^{n} |Wn_{i,j} - n_i|^2 \quad (1)$$

where n_i is the ith neuron at the input layer, $Wn_{i,j}$ is the weight vector connecting input neuron i to output neuron j, and ED_j is Euclidean distance between the input neuron and output neuron, respectively. The winning neuron weight is updated according to the neighborhood function as shown in Eq. 2, and this learning process is repeated until the weights are unchanged.

$$Wn_{i,j}^{q} = Wn_{i,j}^{q-1} + Nh_{i,j}x_i - Wn_{i,j}^{q-1}) \tag{2}$$

where $Wn_{i,j}^{q}$ is new weight vector between neuron i and neuron j. $Wn_{i,j}^{q-1}$ is an old weight vector between i (input layer neuron) and j (output layer neuron). $Nh_{i,j}$ is neighborhood function and x_i is the input sample.

In [7], the authors discuss an Energy-Based Clustering Self-Organization (EBCS) map routing protocol, in which the energy level of nodes and X and Y coordinates of nodes are used as inputs to SOM to make energy balanced clusters. In [8], the author has discussed a routing protocol in which neural networks are used to develop well organized and dynamic topologies to make routing efficient by minimizing the energy utilization in WSNs. In [9], the author has presented an Artificial Neural Network (ANN)-based routing protocol to speed up the lookup table process and make routing easier and efficient in WSNs. The work in [10] discusses a protocol called Adaptive Resonance Theory (ART) neural network that processes high-dimensional complex data into simple low-dimensional data. By doing so, it makes routing efficient in WSNs. Many machine learning-based routing protocols are discussed in [11–16]. Based on this motivation, a SOM-based routing protocol is proposed where we have formed a different number of clusters and identified the optimal number of clusters that balances the energy consumption throughout the network.

3 Proposed Heuristic

The proposed heuristic is developed by incorporating the principle of SOM in WSNs. SOM operates with three inputs such as X coordinates, Y coordinates, and distance to sink node/BS, and the output of SOM is connected to the sensor field. First, the network is partitioned into clusters with the help of winning neurons, and later, the number of clusters is varied with the help of the K-means algorithm. Then, CH is selected based on the centroid of the cluster and the energy level of sensor nodes. The complete procedure is implemented in four phases.

(A) *Initialization Phase*

- Assume sensor nodes are deployed in a specific area. Consider it as N.

- Each sensor node is equipped with the same energy level and identical by nature.
- Each sensor node is static including the base station.

(B) *Cluster formation Phase*

The X and Y coordinates of sensor nodes, and distance from nodes to the BS are used as inputs to SOM. So that a three-dimensional I matrix is created. The values in the input matrix are normalized using the min-max operation.

$$X_{ij}^{j} = \frac{x_{ij} - Min_j}{Max_j - Min_j} \tag{3}$$

In Eq. 3, the X_{ij}^{j} denotes the normalized senor node value of '*j*' column and '*i*' row. The Min_j and Max_j represents the minimum and maximum values of '*j*' column vector. And the x_{ij} denotes the value of '*j*' column and '*i*' row of input vector–matrix '*I*'. The three-dimensional input matrix '*I*' is shown below in Eq. 4.

$$I = \begin{bmatrix} \frac{x_1}{x_{max}} \; \frac{y_1}{y_{max}} \; \frac{d_1}{D_{max}} \\ \cdot \\ \cdot \\ \cdot \\ \frac{x_n}{x_{max}} \; \frac{y_n}{y_{max}} \; \frac{d_n}{d_{max}} \end{bmatrix} \tag{4}$$

In the above input parameters matrix, *I*, the vector $(x_1 \ldots .x_n)$ represents X coordinate values of sensor nodes, the vector $(y_1 \ldots .y_n)$ represents Y coordinates values of sensor nodes, and the vector $(d_1 \ldots .d_n)$ represents the distance from sensor nodes to BS. From the input matrix *I*, the weight matrix is derived as shown below in Eq. 5.

$$WI = \begin{bmatrix} \frac{x_1}{x_{\max}} & \ldots\ldots\ldots & \frac{x_n}{x_{\max}} \\ \frac{y_1}{y_{\max}} & \ldots\ldots\ldots & \frac{y_n}{y_{\max}} \\ \frac{d_1}{d_{\max}} & \ldots\ldots\ldots & \frac{d_n}{d_{\max}} \end{bmatrix} \tag{5}$$

In the above Eq. 5, WI represents the weight matrix of the input matrix I. The X and Y coordinates of sensor nodes and distance to BS are used as input parameters of SOM, fed to the sensor field to form the clusters. The sensor nodes are arranged in grid topological structure and weights of input vectors are randomly initialized. The weights of the input vector are changed according to the learning rate and neighborhood function. The output of SOM is fed to the sensor field, and K-means algorithm is employed to formulate the clusters. The K-means algorithm uses Euclidean distance to make clusters as discussed in Eq. 6.

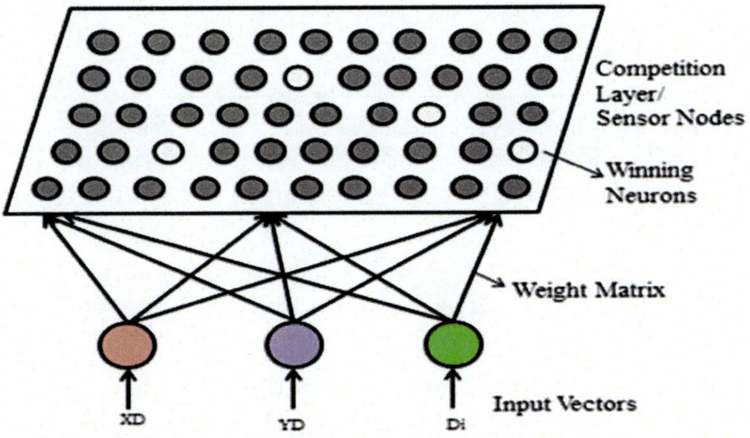

Fig. 3 Proposed WSN clustering model

$$K\text{-means} = \frac{1}{n}\sum_{k=1}^{n}\sum_{x\in C_k}\|x - Cen_k\|^2. \tag{6}$$

In the above equation, '*n*' denotes the number of clusters, C_k identifies the Kth cluster, and Cen_k notifies the centroid of cluster C_k. The proposed model is shown in Fig. 3.

(III) *Cluster Head Selection Phase*
CH is selected after partitioning the network into several clusters. The different number of clusters is formed with the help of the SOM and K-means algorithm. Initially, centroids are found for each cluster, then the CH is selected whose distance to the centroid is low and energy is high. For the entire simulation, the number of clusters remains same, but CHs are rotated based on their energy level.

(IV) *Transmission Phase*
Data transmission phase allows cluster members to transmit the data to their corresponding CHs in each cluster. Then, CHs aggregate the sensor node's data and transmit it to the sink node/BS. For the entire simulation, the clusters remain same, but cluster heads are rotated for each cluster. The energy model is considered the same as the LEACH protocol.

4 Results and Discussion

MATLAB is used to setup the experiments and our findings are graphically characterized from Figs. 4, 5, 6, 7, 8 and 9. Both the LEACH single hop (LEACH-SH) and LEACH multi-hop (LEACH-MH) are taken as the reference protocols to evaluate

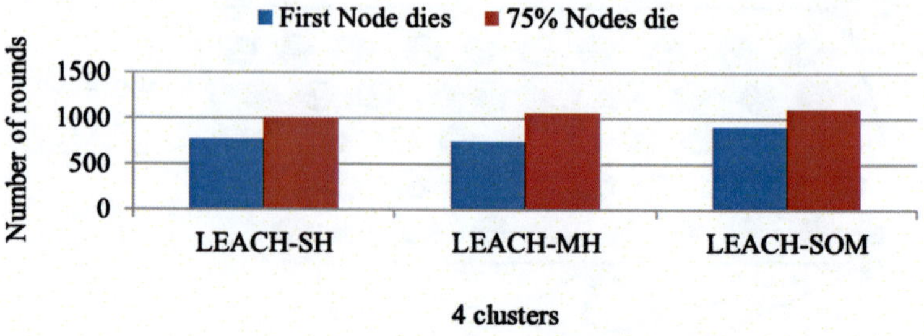

Fig. 4 FND & 75% nodes die w.r.t rounds

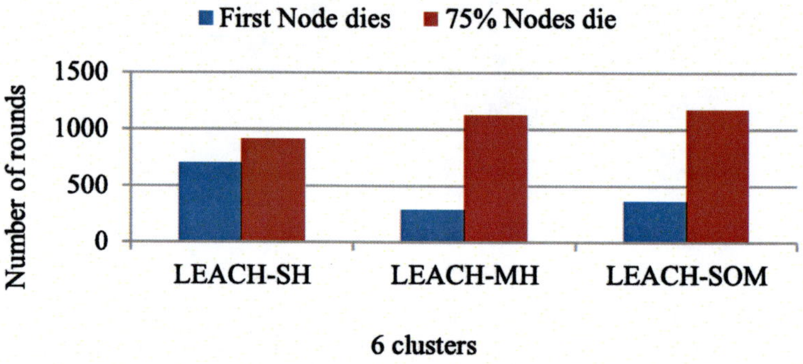

Fig. 5 FND & 75% nodes die w.r.t rounds

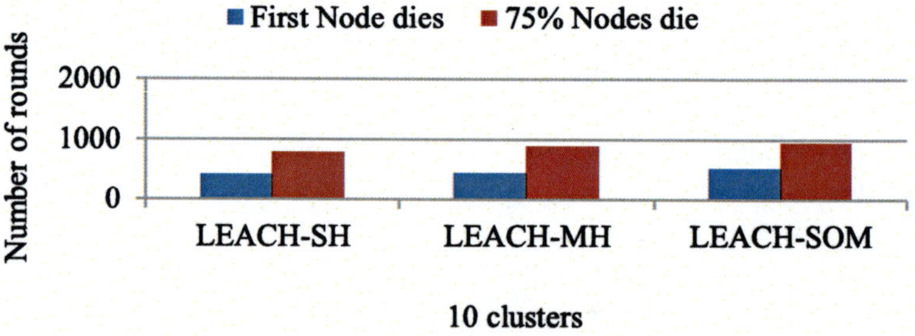

Fig. 6 FND & 75% nodes die w.r.t rounds

the performance of the intended protocol. The simulation parameters are discussed in Table 1. We consider an area of 100*100 m, where 100 sensor nodes are deployed randomly. The base station is placed at 50*200 m. In this approach, the *k* value is varied to 4, 6, and 10 to form 4, 6, and 10 clusters, respectively. Our experimental analysis concludes that 4 clusters are the optimal number of clusters at which the network performs very well compared to 6 and 10 clusters. The performance metrics like energy consumption, packet delivery ratio, and network lifetime are used to

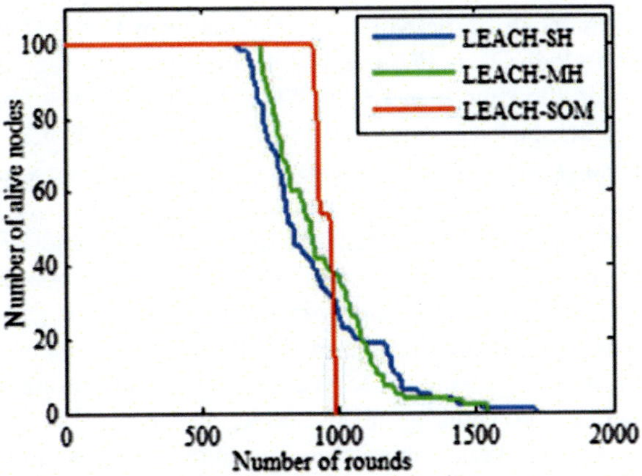

Fig. 7 Network lifetime in terms of alive nodes

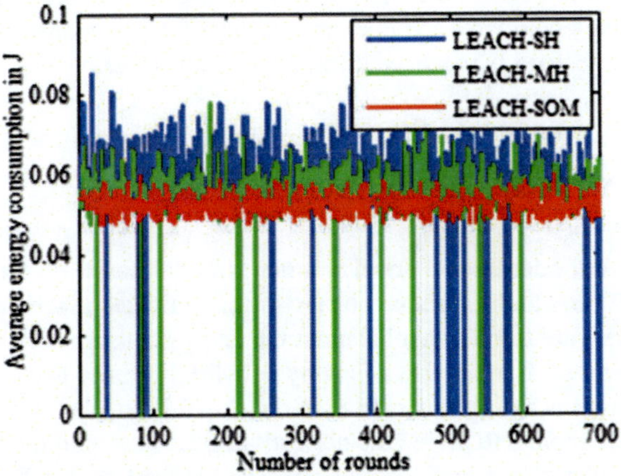

Fig. 8 Average energy consumption versus rounds

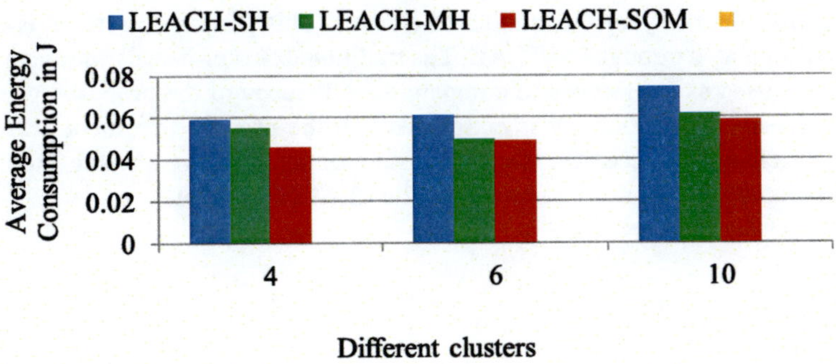

Fig. 9 Average energy consumption versus different size clusters

measure the performance of a network. The figures from 4 to 6 show the time, at what time the First Node Dies (FND) and 75% of nodes die. Figure 4 shows that the first node dies at 896 rounds in LEACH-SOM, whereas First node dies at 664 rounds in LEACH-SH, and it survives up to 638 rounds in LEACH-MH. Usually, the First Node Dies (FND) are the most used performance metric to measure the lifetime of a network. The 75% of nodes are also still alive up to 900 rounds in LEACH-SOM and it is far better compared to LEACH-SH and LEACH-MH. A comparison table about network lifetime in terms of the first node dies, 75% node dies, and the last node dies as shown below in Table 2. Figure 7 is generated through MATLAB which represents the lifetime of the network in terms of the number of alive nodes. We have also generated a MATLAB graph representing average energy consumption per round considering all the three cluster sizes as shown in Fig. 8. The graph shows that the average energy consumption of nodes per round is less in LEACH-SOM related to LEACH-SH and LEACH-MH protocols. Figure 9 depicts the average energy consumption of the network for all sizes of clusters such as four, six, and ten. At cluster size 4, the energy consumption of nodes is less and balanced throughout network compared to other size clusters. So, it is assumed that when the cluster size is large, load is balanced throughout the network.

5 Conclusion

This research work addresses a novel routing protocol based on clustering called "LEACH-SOM" in which effective cluster formation takes place and an accurate number of clusters are formed that balances the energy consumption throughout the network. By doing so, it maintains a healthy network lifetime. The Self-Organizing Map and the K-means algorithm are used to develop this novel cluster-based routing protocol. SOM is applied to divide the network into clusters, and the K-means algorithm is used to vary the size of the clusters. The parameters like X coordinates, Y coordinates, and distance to BS of sensor nodes are fed to the input of the SOM, and the output of SOM is fed to the sensor field. K-means algorithm is applied to the sensor field to vary the number of clusters. This proposed protocol is examined and simulated through MATLAB. The traditional cluster-based routing protocol LEACH is used as a reference to compare the efficiency of the proposed protocol. After extensive simulations, it is concluded that the proposed protocol "LEACH-SOM" efficiently balances the energy consumption and brings a stable lifetime of Wireless Sensor Network as compared to the LEACH protocol.

Table 1 Simulation Parameters

Sensor nodes	100
Simulation terrain	100 × 100 m
Sink or BS	50 × 200 m
$d_{crossover}$	87 m
Initial energy of sensor nodes	0.5j
E_{elec}	50nj/bit
E_{mp}	0.0013pj/bit/m^4
Efs	10pj/bit/m^2
Data packet length	500 bytes
EDA	5nj/bit/signal
Speed of transmission	1 Mbps
Channel representation	Wireless (both directions)
Simulation period	1001 s

Table 2 Time of dead nodes w.r.t 4 clusters

Protocol name	Time in terms of no. of rounds		
	1st node dies	75% nodes die	Last node dies
LEACH-SH	664	836	1801
LEACH-MH	638	888	1533
LEACH-SOM	896	990	1010

References

1. Heinzelman, W. R., Chandrakasan, A., & Balakrishnan, H. (2000). Energy-efficient communication protocol for wireless microsensor networks. *IEEE Computer Society Proceedings of the Thirty Third Hawaii International Conference on System Sciences (HICSS '00). IEEE, Washington, DC, 2000* (pp. 1–10)
2. Heinzelman, W. R., Chandrakasan, A., Balakrishnan, H. (2002). An application-specific protocol architecture for wireless microsensor networks. *IEEE Transactions on Wireless Communications, 1*(4), 660–670.
3. Lindsey, S., & Raghavendra, C. S. (2002). PEGASIS: Power efficient gathering in sensor information systems. *IEEE Aerospace Conference Proceedings*
4. Manjeshwar, A., & Agrawal, D. P. (2001). TEEN: A routing protocol for enhanced efficiency in wireless sensor networks. *Proceedings of the 15th International Parallel & Distributed Processing Symposium (IPDPS-01), San Francisco, CA, April 23–27.*
5. Manjeshwar, A., & Agrawal, D. P. (2002). APTEEN: A hybrid protocol for efficient routing and comprehensive information retrieval in wireless sensor networks. *16th International Parallel and Distributed Processing Symposium (IPDPS 2002).*
6. Priyadarshi, R. et al. (2018). A novel HEED protocol for wireless sensor networks. *5th International Conference on Signal Processing and Integrated Networks (SPIN).*
7. Enami, N., Moghadam, R. A., & Ahmadi, K. D. (2010). A new neural network based energy efficient clustering protocol for wireless sensor networks. *IEEE 5th International Conference on Computer Sciences and Convergence Information Technology.*

8. Patra, C., Bhaumik, P., Chattopadhyay, M., & Roy, A. G. (2011). Using self organizing map in wireless sensor network for designing energy-efficient topologies. *IEEE 2011 2nd International Conference on Wireless Communication, Vehicular Technology, Information Theory and Aerospace & Electronic Systems Technology (Wireless VITAE).*
9. Turčaník, M. (2012) Network routing by artificial neural network. *IEEE 2012 Military Communications and Information Systems Conference (MCC).*
10. Mittal, M., & Kumar, K. (2012) Network lifetime enhancement of homogeneous sensor network using ART1 neural network. *IEEE Sixth International Conference on Computational Intelligence and Communication Networks.*
11. Li, L., & Liu, D. (2016). An energy aware distributed clustering routing protocol for energy harvesting wireless sensor networks. *2016 IEEE/CIC International Conference on Communications in China (ICCC).*
12. Zhao, X., Wei, Z., Cong, Y., Yin, B. (2018). A balances energy consumption clustering routing protocol for a wireless sensor network. *IEEE 4th Information Technology and Mechatronics Engineering Conference (ITOEC 2018).*
13. Sanhaji, F., Satori, H., Satori, K. (2019). Cluster head selection based on neural networks in wireless sensor networks. *IEEE 2019 International Conference on Wireless Technologies Embedded and Intelligent Systems (WITs).*
14. Nayak, P., & Anurag, D. (2016). Distributed fuzzy logic based routing protocol for wireless sensor networks. *IEEE Sensor Journal,* 16(1), 137–144.
15. Nayak, P., & Bhavani, V. (2017). Energy efficient clustering algorithm for multi-hop wireless sensor network using type-2 fuzzy logic. *IEEE Sensor Journal, 17*(14), 4492–4499.
16. Swami, T. J., Ramamurthy, G., & Nayak, P. (2019). Optimal, secure cluster head placement through source coding techniques in wireless sensor networks. *IEEE Communication Letters, 24*(2), 443–446.

CNN-Based Mobile Device Detection Using Still Images

Surbhi Gupta, Neela Pravalika, Padmalaya Nayak, and Jaafar Al Ghazo

Abstract Image forgery has been increased enormously due to the development of software and introduction of new cheaper digital devices like cameras and cellular phones. The introduction of new devices leads to inadvertent capturing of images. A lot of misuse of these devices led to the need for their identification. Source camera identification deals with the issue of identification of the mobile device using their camera through which image has been taken. This empowers the scientific agent to spot mobile device that has been used for capturing the specific image during the study. This is significant as this alphanumeric content is considered as a inaudible observer. In this proposed work, we are reviewing the various approaches for SCI which depend on classical machine learning algorithms feature extraction. Then, a CNN model is proposed for identification of mobile device using vision dataset. High accuracy of 92% is achieved for vision dataset.

Keywords Source camera identification · Convolutional neural network · Sensor fingerprint · Feature extraction · Mobile devices · Classification

1 Introduction

Digital images can be captured through digital cameras or mobile devices. The authentication of questioned smartphone images is very important [20]. Image forgery has been increased rapidly with the growth of digital technologies. Now the pictures can be easily captured without any constraints of network, location, time, etc. Lately, a large amount of multimedia data is produced with various distinct electronic gadgets and exchanged through the internet [7]. This also led to enforcing legal restrictions of capturing the pictures which include places like schools, government places [9]. The two main important domains in multimedia forensics are forgery

S. Gupta · N. Pravalika · P. Nayak
Gokaraju Rangaraju Institute of Engineering and Technology, Hyderabad, Telangana, India

J. Al Ghazo (✉)
Prince Mohammad Bin Fahd University, Khobar, Saudi Arabia

P. Nayak et al. (eds.), *IoT and Analytics for Sensor Networks*, Lecture Notes in Networks and Systems 244, https://doi.org/10.1007/978-981-16-2919-8_3

detection and camera documentation [32]. Forgery detection has gained huge attention, and it can be achieved by the identification of source camera. It has many useful applications in different areas like a criminal investigation, the judicial system, military image applications, and news media, etc. Digital images that are captured through cameras, mobile phones, tablets, etc., can be used for illegitimate purposes. A digital image may act as a "proof of occurrence" of an incident [16]. Many times, after a crime, a forensic officer is required to authenticate the images which may be considered as proof. For this purpose, the image source is first identified. The fixing of the image source gives a clue about whether the content of images has tampered or not. Source camera identification and its usage have become very prominent in the present time [33]. It is the process of identification of the mobile device which is used to capture the image. It is considered as evidence in legal terminology. It emphasizes the relationship between mobile cameras and pictures. The foremost aim of this paper is to conduct an assessment of SCI and then propose a better model.

2 Techniques for Source Camera Identification

The main practices to tackle the problem of identification of source cameras for mobile devices are using image headers, sensor fingerprint extraction, and classification, deep learning, etc. Gupta et al. [13] have presented a complete survey of source camera attribution for camera devices. In this paper, we aim at detecting camera device using still images.

2.1 Image Headers

Mullan et al. [24] used JPEG headers and metadata for source camera identification. Digital image metadata is very important to maintain digital image repositories [28]. The image header information includes metadata like parameterization and Exchangeable Image File Format (EXIF) tags. The EXIF metadata recorded by the camera improves the classification accuracy [6]. In this work, source camera identification was done using metadata from an image container. The ordering of the iPhone Model through quantization matrices and EXIF directories was done and accuracy is 65.6% is obtained.

2.2 Images Features

Tsai et al. [30] presented a novel practice for the documentation of source camera for mobile devices. It uses a set of image features for finding the different features of a mobile camera. These features comprise color, quality, and image features and characteristics of the frequency domain. Numerical image structures also show an prominent part in source camera documentation [15, 22]. The wavelet transform is studied to calculate wavelet domain statistics-based structures. These are categorized using a Support Vector Machine (SVM). They attained accuracy close to 92% for 4 camera models for two different camera brands. McKay et al. [19] stretched the identification of foundation device categories like digital scanners, phone cameras, and mobile cameras, etc.

2.3 Demosaicing Features

Image demosaicing is a process of interpolating a full resolution colored image from Color Filter Array (CFA) samples. In an initial work [19], identification of the source camera model is done blindly by using three types of demosaicing features derived from the given text picture. It is shown that the Eigen demosaicing features function performs remarkably better than other regular features. Nine portable cameras of different prototypes are differentiated reliant upon collected image chunks through Eigen feature regularization and feature reduction [8]. For the given input image, this framework first estimates the demosaicing weights then extracts the demosaicing feature and then regularizes and compresses Eigen features.

2.4 PRNU Feature-Based Approach

Photo Response Non-Uniformity (PRNU) noise is a important technique for SCI. The sensor fingerprint is an exclusive characteristic that is useful for differentiating images taken from the same type of cameras. PRNU has been proved as an effective approach for digital forensics. The drawback of the earlier PRNU-based models [25] is that they are sensitive to random noise. These are not capable of identifying any manipulation or changes in the image or picture. So, a new feature-based PRNU technique for identification of source camera for mobile devices is devised by [14] which chooses the features from the images. These are robust for performing any simple geometric variations or any manipulations on the image such as rotation and scaling of the image. By using the wavelet-based denoising method [1], PRNU noise is obtained from images. This is represented by Higher-Order Wavelet Statistics (HOWS), and this is invariable characteristics for any kind of symmetrical dissimilarities or image manipulations in the image.

2.5 Mobile Sensor Fingerprint

Mobile sensor fingerprints demonstrate how several sensors on a smartphone can be utilized to build good hardware pattern of the mobile device. This pattern could be utilized to a identify a mobile device as they link to the websites. The accelerometer-based pattern is very fascinating as it is approachable through JavaScript processing in a web browser without the invocation of any authorisation of the user [5]. Several experiments have been conducted by using sensor fingerprint-based properties from ten thousand mobile devices. It is shown that the entropy from sensor fingerprint is solely required to recognize devices from various other devices, with less chance of collision [5].

2.6 Convolutional Neural Networks and Image Noise Pattern

Zuo et al. [34] introduced this technique. The dataset from Kaggle competition, named "Camera Model Identification", is used. CNNs can involuntarily and parallelly extract features and learn to classify during the learning process [31]. It is proved that CNN-based architecture could be utilized for solving the problem of camera detection. In this work, the author proposed a CNN classifier after performing PRNU extraction. PRNU is the most known SCI method in digital forensics [23]. This model is not very good as it is a simple CNN model. Here, only two filters have been used, and the number of epochs used is only 100 for testing purposes. It is difficult to estimate the performance of the model by using a greater number of epochs or filters. In the future, more experiments are to be expected to conclude. The accuracy obtained using this method is 91%. Deep learning depends upon the input data quality for its feature extraction process [11]. Lately, Convolutional Neural Networks (CNNs) have exhibited incredible performance on various projects which include video analysis, image recognition, etc. A CNN comprises a collection of layers in which every layer is an association of different operations or tasks. Mobile camera sensors consist of a noise pattern that can be easily understood by a CNN architecture [18]. This is as well as called camera fingerprint which is used to identify the source mobile device through which image has been captured with good accuracy. The fingerprint is mainly used for two purposes, i.e., smartphone verification and profile linking [4]. PRNU is mostly used for identifying sensor fingerprint [2]. In most of the existing systems using classifier or neural networks, firstly, the features are retrieved from the given image. The dataset is subdivided into testing and training data. Then, a classifier model is obtained through training and evaluation of the model is done through testing. In this paper, we studied the various approaches that have been utilized for the identification of source camera for mobile devices. The several techniques that have been discussed for the identification of source cameras are using image headers, image features, demosiacing features, sensor fingerprint, etc. These are reviewed individually. The comparison of different methods is shown in Table 1.

Table 1 Evaluation of source camera identification techniques for mobile device

Author	Algorithm	Year	Format of image	Number of brands	Number of models	Applied to mobiles	Applied to same brand and different models	Accuracy
Tsai et al. [30]	SVM	2007	JPEG	2	4	Yes	Yes	92%
Cao and Kot [8]	Eigen Feature Regularization	2010	JPEG	4	15	Yes	Yes	99%
Akshatha et al. [1]	SVM	2016	JPEG	7	11	Yes	Yes	96.18%
Zuo et al. [34]	CNN + PRNU	2018	JPEG	3	3	Yes	No	91%
Mullan et al. [1]	JPEG Image Headers	2019	JPEG	1	7	Yes	No	81.70%
Freire-Obregón et al. [11]	CNN	2018	JPEG	2	3	Yes	Yes	98.10%

3 Dataset Used

A dataset is a very important aspect while performing any research-related experiment. There are different types of datasets that are used for source camera identification for mobile devices. Some people create their own dataset while other people prefer using existing datasets. The dataset that we have considered for the proposed model is the VISION dataset. Shullani et al. [27] introduced dataset in which images are taken from several different mobile devices, and it also consists of the images shared from WhatsApp and Facebook. It is a development to be used for mobile forensics. This dataset consists of 34,427 images and 1914 videos captured using 35 mobile devices. Some of them are shown in Fig. 1. Before starting the experiment, the dataset is split as 80% of total data for training and validation and remaining data for testing. Table 2 shows the 11 mobile devices that we have considered from the VISION dataset.

4 Proposed CNN Model and Its Architecture

Convolutional neural networks do not differ much from the other neural network. CNN could generate its own features. It consists of neurons. These are like the neurons in biology. Each neuron has its own parameters, i.e., height and width. Every neuron in the network has a specific operation to perform. These operations differ from one neuron to the other neuron. A CNN is a neural network in which given an input, then it performs a set of operations on the input. It also consists of non-linearity operation, i.e., an activation function for every computational process. At the end of

Fig. 1 Images from VISION dataset

Table 2 Dataset with 11 mobile devices from VISION dataset

S. No	Name of the device	Total images
1	Samsung_GalaxyS3Mini	901
2	Apple_iPhone4S	919
3	Huawei_p9	1066
4	LG_D290	1044
5	Lenovo_P70	809
6	Sony_XperiaZ1Compact	964
7	Microsoft_Lumia640LTE	894
8	WikoRidge4g	2912
9	Asus-Zenfone 2 lazer	957
10	Xiaomi_RedmiNote3	1422
11	OnePlus_A3	1324

the process, it is feed to a loss function. It plays a prominent role in every neuron parameter apprise. CNN considers only image as an input. The idea behind the CNN is to reduce the number of parameters. The different layers that are present in CNN architecture are convolutional, pooling, and fully connected layer. The architecture for the model should be fixed before the start of training model as shown in Fig. 2.

5 Results and Discussions

The accuracy obtained for these 11 mobile device datasets is 92.39%. The accuracy graph obtained for 11 mobile device datasets is shown in Fig. 3. As we can see the

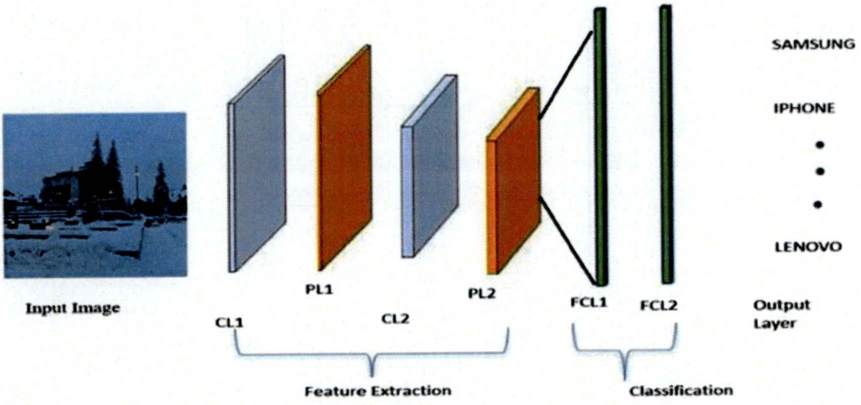

Fig. 2 Architecture for the proposed model

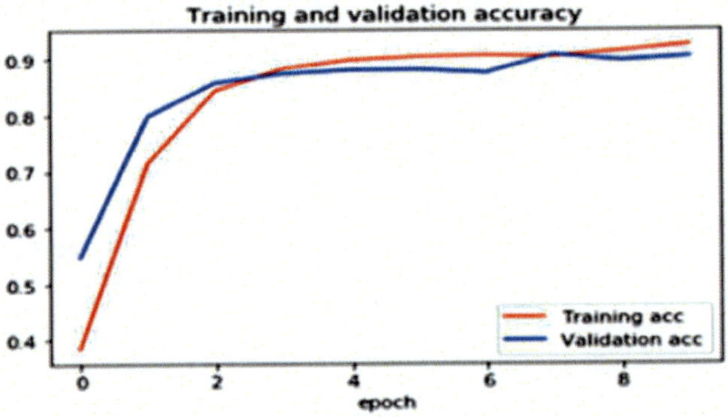

Fig. 3 Graphs obtained for 11 mobile device dataset

validation accuracy is following the trend for training accuracy and is rising after 2nd epoch itself. The experiment has been conducted only for 10 epochs. More epochs will help in improving the accuracy. The confusion matrix for 11 devices has been showed in Table 3. It is evident that almost all the models are correctly classified with 0.99 precision except 'Asus' and 'Wiko'. This is the reason that the overall accuracy dropped by 8%.

Table 4 illustrates the contrast of recommended method with the present methods. Most of the existing techniques used Miche dataset. But this dataset is limited as only three different device images are included. We have used Vision dataset with 11 model. Existing shallow CNN [21] has achieved 95% but we have proposed a lighter CNN with 92.39% accuracy. Moreover, the epochs used are only 10.

Table 3 Confusion matrix for 11 mobile device dataset

	Apple	Asus	Oneplus	Huawei	Lenovo	LG	Microsoft	Redmi	Samsung	Sony	Wiko
Apple	900	0	0	4	2	0	1	4	0	0	8
Asus	0	22	0	0	0	0	0	0	0	0	935
Oneplus	0	0	1320	0	4	0	0	0	0	0	0
Huawei	0	0	1	1065	0	0	0	0	0	0	0
Lenovo	0	0	0	0	809	0	0	0	0	0	1
LG	0	0	0	7	0	1040	0	1	0	0	0
Microsoft	0	0	0	0	4	0	843	2	0	0	0
Redmi	4	0	0	0	0	0	0	1418	0	0	0
Samsung	0	0	10	2	1	1	1	1	811	3	19
Sony	2	0	0	0	0	0	1	0	0	961	0
Wiko	0	22	0	0	0	0	4	4	6	0	2876

Table 4 Contrast with present state of art

Method proposed	Dataset	Image input size	No of the CNN layers employed	No of devices experimented	Achieved accuracy
Barrofio et al. [3]	Miche-I	32 × 32	7	3	92.3%
Tuama et al. [31]	Miche-I	32 × 32	7	3	93.4%
Freire-Obregón et al. [11]	Miche-I	32 × 32	6	3	98.1%
Shallow CNN [21]	Vision	64 × 64	5	11	95.06%
Proposed shallow CNN	Vision	150 × 150	4	11	92.39%

6 Conclusion

In this paper, a complete survey of different approaches used for source mobile detection is analyzed. Most of the techniques for the identification of source cameras are using image headers, image features, Demosiacing features, sensor fingerprint, etc. These are reviewed individually. Although these are good at the discrimination of the mobile model, only some methods succeeded in identifying the sensor level camera. More work is required in this direction. CNN model is proposed for identification of mobile device using vision dataset. High accuracy is achieved. In future, ensemble CNN archirecture-based SCI technique can be proposed for better accuracy. The same model can be explored for other computer vision applications [12, 17].

References

1. Akshatha, K. R., Karunakar, A. K., Anitha, H., Raghavendra, U., & Shetty, D. (2016). Digital camera identification using PRNU: A feature based approach. *Digital Investigation, 19*, 69–77.
2. Amerini, I., Caldelli, R., Cappellini, V., Picchioni, F., & Piva, A. (2009). Analysis of denoising filters for photo response non uniformity noise extraction in source camera identification. In 2009 *16th International Conference on Digital Signal Processing* (pp. 1–7). IEEE.
3. Baroffio, L., Bondi, L., Bestagini, P., Tubaro, S. (2016). Camera identification with deep convolutional networks, *CoRR* abs/1603.01068.
4. Bertini, F., Sharma, R., Iannì, A., & Montesi, D. (2015). Smartphone verification and user profiles linking across social networks by camera fingerprinting. In *International Conference on Digital Forensics and Cyber Crime* (pp. 176–186). Springer, Cham.
5. Bojinov, H., Michalevsky, Y., Nakibly, G., & Boneh, D. (2014). Mobile device identification via sensor fingerprinting. arXiv preprint arXiv:1408.1416.
6. Boutell, M., & Luo, J. (2004). Photo classification by integrating image content and camera metadata. In *Proceedings of the 17th International Conference on Pattern Recognition*, ICPR 2004., Vol. 4, pp. 901–904. IEEE.
7. Caldelli, R., Becarelli, R., & Amerini, I. (2017). Image origin classification based on social network provenance. *IEEE Transactions on Information Forensics and Security, 12*(6), 1299–1308.

8. Cao, H., & Kot, A. C, 2010, May. Mobile camera identification using demosaicing features. In *Proceedings of 2010 IEEE International Symposium on Circuits and Systems* (pp. 1683–1686). IEEE.
9. Corripio, J. R., González, D. A., Orozco, A. S., Villalba, L. G., Hernandez-Castro, J., & Gibson, S. J. (2013). Source smartphone identification using sensor pattern noise and wavelet transform
10. De Marsico, M., Nappi, M., Riccio, D., & Wechsler, H. (2015). Mobile iris challenge evaluation (MICHE)-I, biometric iris dataset and protocols. *Pattern Recognition Letters, 57*, 17–23.
11. Freire-Obregón, D., Narducci, F., Barra, S., & Castrillón-Santana, M. (2019). Deep learning for source camera identification on mobile devices. *Pattern Recognition Letters, 126*, 86–91.
12. Gupta, S., & Kumar, M. (2020). Forensic document examination system using boosting and bagging methodologies. *Soft Computing, 24*(7), 5409–5426.
13. Gupta, S., Mohan, N. & Kumar, M. A. (2020). Study on source device attribution using still images. *Arch Computat Methods Eng* (2020). https://doi.org/10.1007/s11831-020-09452-y.
14. Darvish Morshedi Hosseini, M., & Goljan, M. (2019). Camera identification from HDR images. In *Proceedings of the ACM Workshop on Information Hiding and Multimedia Security* (pp. 69–76).
15. Hu, Y., Li, C. T., & Zhou, C. (2010) Selecting forensic features for robust source camera identification. In *2010 International Computer Symposium* (ICS2010) (pp. 506–511). IEEE.
16. Kharrazi, M., Sencar, H. T., & Memon, N. (2004). Blind source camera identification. In *2004 International Conference on Image Processing*, ICIP'04., Vol. 1, pp. 709–712). IEEE.
17. Kumar, M., Gupta, S., Gao, X. Z., & Singh, A. (2019). Plant species recognition using morphological features and adaptive boosting methodology. *IEEE Access, 7*, 163912–163918.
18. Lawgaly, A., Khelifi, F., & Bouridane, A. (2014). Weighted averaging-based sensor pattern noise estimation for source camera identification. In *2014 IEEE International Conference on Image Processing* (ICIP) (pp. 5357–5361). IEEE.
19. Li, X., Gunturk, B., & Zhang, L. (2008) Image demosaicing: A systematic survey. In *Visual Communications and Image Processing* 2008 (Vol. 6822, p. 68221J). International Society for Optics and Photonics.
20. Liu, Q., Li, X., Chen, L., Cho, H., Cooper, P. A., Chen, Z., Sung, A. H. (2012). Identification of smartphone-image source and manipulation. In *International Conference on Industrial, Engineering and Other Applications of Applied Intelligent Systems* (pp. 262–271). Springer, Berlin, Heidelberg.
21. Marra, F., Gragnaniello, D., & Verdoliva, L. (2018). On the vulnerability of deep learning to adversarial attacks for camera model identification. *Signal Processing: Image Communication, 65*, 240–248.
22. McKay, C., Swaminathan, A., Gou, H., & Wu, M. (2008) Image acquisition forensics: Forensic analysis to identify imaging source. In *2008 IEEE International Conference on Acoustics, Speech and Signal Processing* (pp. 1657–1660). IEEE.
23. Mohanty, M., Zhang, M., Asghar, M. R., & Russello, G. (2018). PANDORA: Preserving privacy in PRNU-based source camera attribution. In *2018 17th IEEE International Conference On Trust, Security And Privacy In Computing And Communications/12th IEEE International Conference On Big Data Science And Engineering* (TrustCom/BigDataSE) (pp. 1202–1207). IEEE.
24. Mullan, P., Riess, C., & Freiling, F. (2019). Forensic source identification using JPEG image headers: The case of smartphones. *Digital Investigation, 28*, S68–S76.
25. Peng, F., Shi, J., & Long, M. (2013). Comparison and analysis of the performance of PRNU extraction methods in source camera identification. *The Journal of Computer Information Systems, 9*(14), 5585–5592.
26. Shaya, O. A., Yang, P., Ni, R., Zhao, Y., & Piva, A. (2018). A new dataset for source identification of high dynamic range images. *Sensors, 18*(11), 3801.
27. Shullani, D., Fontani, M., Iuliani, M., Al Shaya, O., & Piva, A. (2017). Vision: A video and image dataset for source identification. *EURASIP Journal on Information Security, 2017*(1), 15.
28. Tesic, J. (2005). Metadata practices for consumer photos. *IEEE Multimedia, 12*(3), 86–92.

29. Tian, H., Xiao, Y., Cao, G., Zhang, Y., Xu, Z., & Zhao, Y. (2019). Daxing smartphone identification dataset. *IEEE. Access, 7*, 101046–101053.
30. Tsai, M. J., Lai, C. L., & Liu, J. (2007) Camera/mobile phone source identification for digital forensics. In *2007 IEEE International Conference on Acoustics, Speech and Signal Processing-ICASSP'07*, Vol. 2, pp. II-221. IEEE.
31. Tuama, A., Comby, F., & Chaumont, M. (2016). Camera model identification with the use of deep convolutional neural networks. In *2016 IEEE International workshop on information forensics and security* (WIFS) (pp. 1–6). IEEE.
32. Van Lanh, T., Chong, K. S., Emmanuel, S., & Kankanhalli, M. S. (2007). A survey on digital camera image forensic methods. In *2007 IEEE international conference on multimedia and expo* (pp. 16–19). IEEE.
33. Zeng, H., & Kang, X. (2016). Fast source camera identification using content adaptive guided image filter. *Journal of forensic sciences, 61*(2), 520–526.
34. Zuo, Z. (2018). Camera model identification with convolutional neural networks and image noise pattern.

E-FFTF: An Extended Framework for Flexible Fault Tolerance in Cloud

Moin Hasan, Major Singh Goraya, and Tanya Garg

Abstract Fault tolerance provisioning is extremely important in cloud. Literature suggests that the existing frameworks are service provider centric. Being a pay-per-use model, fault tolerance in the cloud should be flexible with respect to the users' requirement. Our previously proposed framework FFTF (Flexible Fault Tolerance Framework) is further extended to E-FFTF in the present paper to provide user transparency in the selection of service category based on the task completion deadline and slack time. E-FFTF successfully establishes task execution price savings through flexible fault tolerance. E-FFTF is proved to be beneficial for both the cloud service providers and service users as per the obtained experimental results.

Keywords Cloud computing · Fault tolerance · Flexible fault tolerance · Reliability · Service price saving

1 Introduction

Fault tolerance is among the most imperative issues in cloud due to its dynamic architecture [1, 2]. Cloud follows the service-oriented architecture where computing services are provisioned on demand basis in a pay-per-use manner [3]. Therefore, provisioning fault tolerance in a flexible manner where users can avail it as per their budgetary constraints will consequently strengthen the service orientation of cloud.

In our previous research work [4], FFTF (flexible fault tolerance framework) is proposed in cloud using cooperative resource group based on the popular replication technique [5, 6]. Replicated resource groups are organized with respect to the

M. Hasan (✉)
Lovely Professional University, Phagwara, India
e-mail: moin.25676@lpu.co.in

M. S. Goraya
Sant Longowal Institute of Engineering and Technology, Sangrur, India

T. Garg
Thapar Institute of Engineering & Technology, Patiala, India
e-mail: tanya.garg@thapar.edu

P. Nayak et al. (eds.), *IoT and Analytics for Sensor Networks*, Lecture Notes in Networks and Systems 244, https://doi.org/10.1007/978-981-16-2919-8_4

quantized reliability of the cloud hosts quantized into five fuzzy sets. Flexible fault tolerance is then realized through three service categories (premium, standard, and economy) by organizing the replicated groups from these cloud hosts. Premium category services are highly fault-tolerant, suitable for the users with strict task deadlines. Standard and economy category services offer relatively lower fault tolerance. They are suitable for the users with softer task deadlines. Although, FFTF achieves flexibility in fault tolerance, still following are the important issues which needs further exploration: (a) Service selection is largely user centric where users select the service as per their budgetary constraints. In such case, it is possible that inexperienced users may choose inappropriate service category. (b) FFTF does not give an insight into price saving through flexible fault tolerance. In E-FFTF, we endeavour to address the above issues with the following contributions:

- *Developing a service selection scheme*: We propose three service classes (A, B, and C) in the decreasing order of reliability. E-FFTF selects suitable class for the users corresponding to the criticality of their tasks based on their respective deadlines. Slack time of a task is used to define the level of fault tolerance.
- *Price saving evaluation through flexible fault tolerance*: Reliability is an important parameter of fault tolerance [7]. The prominent cloud providers claim their reliability above 99 percent.[1] Taking this as benchmark, we evaluate the final price saving through flexible fault tolerance in E-FFTF.

E-FFTF is beneficial for both the provider as well as the user. Service provider can extend the low reliability resources to the tasks with low level of fault tolerance. Thus, the low reliability resources may not be instantly required to be pushed to the obsolete state. Cloud users will have the chance of price saving for the extended task deadline.

In rest of the paper, Sect. 2 describes the proposed E-FFTF explaining the system model and task execution model. Experimental evaluation is given in Sect. 3 and the paper is finally concluded in Sect. 4.

2 Proposed E-FFTF

This Section describes E-FFTF by giving an insight into the system model, fault model, and task execution model.

2.1 System Model

Cloud data center is considered as a set $DC = \{PM_1, PM_2, PM_3, \ldots, PM_m\}$ of m physical machines (PMs). PM_i in the DC is modelled by its processing power

[1] https://www.enterpriseai.news/2015/01/06/aws-rates-highest-cloud-reliability/.

$\left(PM_i^{pp}\right)$ evaluated in Million Instructions per Second (MIPS), working state $\left(PM_i^{ws}\right)$ which can be failed or repaired, operational mode $\left(PM_i^{om}\right)$ which can be active or sleep, and quantized reliability $\left(PM_i^{qr}\right)$. The raw reliability of PM_i for time x is calculated as a single module repairable system through Gamma distribution as

$$PM_i^{rr}(x) = \int_x^{\infty} \frac{\beta^{\alpha} x^{\alpha-1}}{\Gamma(\alpha)} e^{-x\beta} dx \tag{1}$$

Reliability of PMs is quantified as Very Low $(VL = 0.1)$, Slightly Low $(SL = 0.3)$, Moderate $(MD = 0.5)$, Slightly High $(SH = 0.7)$, Very High $(VH = 0.9)$ using fuzzy logic [4]. $PM_i = \{VM_{i1}, VM_{i2}, VM_{i3}, \ldots, VM_{in}\}$ represents n number of virtual machines (VMs) deployed on demand over PM_i. VM_{ij} is modelled by its processing power $\left(VM_{ij}^{pp}\right)$ evaluated in MIPS and quantized reliability VM_{ij}^{qr} considered equivalent to PM_i^{qr}.

PMs observe failures due to various faults in cloud. Multiple PMs may fail simultaneously, but independently. Failure of VMs are considered totally dependent on their respective parent PMs. All the VMs deployed over PM_i will be automatically undeployed at the event PM_i fails in the data center [8].

2.2 *Task Execution Model*

A finite set of p independent tasks $T = \{t_1, t_2, t_3, \ldots, t_p\}$ is considered for execution [4]. Tasks are broadly categorized as delay-sensitive and delay-insensitive. Delay-sensitive tasks are high priority tasks with some specific completion deadline and require a strong fault tolerance mechanism. A delay-sensitive task t_k is modelled by its arrival time $\left(t_k^{at}\right)$, task size $\left(t_k^{ts}\right)$ in Million Instructions (MI), deadline $\left(t_k^{dl}\right)$, and service class $\left(t_k^{sc}\right)$. Delay-insensitive are the low priority tasks with no specific deadline. These tasks improve the resource utilization of the system. A delay-insensitive task t_k is modelled by its arrival time $\left(t_k^{at}\right)$ and task size $\left(t_k^{ts}\right)$. If t_k is allocated for execution to VM_{ij}, its execution time t_k^{et} and expected finish time t_k^{ft} are calculated as

$$t_k^{et} = \frac{t_k^{ts}}{VM_{ij}^{pp}} \tag{2}$$

$$t_k^{ft} = t_k^{st} + t_k^{et} \tag{3}$$

where, t_k^{st} is the starting time of the task execution. If the deadline of a delay-sensitive task falls beyond its expected finish time based on the system performance, then its slack time (t_k^{sl}) is evaluated as

Table 1 Service Classes and their reliability

Service Class	Reliability
Class A	0.962
Class B	0.814
Class C	0.624

$$t_k^{sl} = t_k^{dl} - t_k^{ft} \tag{4}$$

Slack time is an important parameter while selecting the service. Large value of slack time corresponds to softer deadline where the task can be executed in a low reliability service class resulting in the price saving.

To execute a delay-sensitive task t_k, replication group $G_k = \{VM_{i_1j_1}, VM_{i_2j_2}, VM_{i_3j_3}\}$, consisting of three VMs deployed over different PMs, is organized in the data center based on the quantified reliability of VMs. Among quantified reliability values, three VMs can be selected in thirty-five ways [9]. Each replication group is considered as a multi-module repairable system where the reliability is calculated as

$$Rel(G_k) = 1 - \prod_{y=1}^{3}\{1 - VM_{i_yj_y}^{qr}\} \tag{5}$$

Thirty-five combinations are clustered further into classes (A, B, and C) using k-means clustering [10]. Class A represents high, Class B represents moderate, and Class C represents relatively lower fault tolerance. Table 1 enlists the three service classes. Reliability lesser than 0.5 is neglected while realizing the service classes.

Corresponding to the slack time, group G_k is organized for t_k in a suitable service category as follows:

- Let t_k^{sl} be the slack time of task t_k, then the execution time in terms of its slack time and deadline will be given as $t_k^{et} = t_k^{dl} - t_k^{sl}$.
- For each service category (in the increasing order of reliability), if ($t_k^{dl} \times$ *serice reliability* $> t_k^{et}$), organize the G_k in the current service category.

For a task t_k, the price of accessing a service is calculated as follows: Let the price of provisioning 250$MIPS$ for 100$seconds$ with reliability approaching to 1 be the base price equal to 1$unit$. Then, to execute t_k for t_k^{dl} seconds on VM_{ij} in G_k, the execution price t_k^{ep} is evaluated as

$$t_k^{ep} = \left(VM_{ij}^{pp} \times t_k^{dl} \times Rel(G_k)\right)/25000 \tag{6}$$

After the group organization, delay-sensitive task execution is started on the maximum reliability VM (nominated as primary) in the group. The remaining VMs in the group are nominated as backups.

On the other hand, to execute a delay-insensitive task t_k, an idle backup VM in the existing replication groups is searched to initiate the task. Delay-insensitive tasks' execution on the backup VMs considerably increases the resource utilization of the replication groups.

Tasks' execution states are periodically checkpointed, so that minimum rollback would be required in the event of a failure. Let the time interval between two successive checkpoint operations be denoted by Int_{CP}, then the updating in the task size t_k^{ts} of the task t_k allocated to VM_{ij} after a checkpoint interval is given as follows:

$$t_k^{ts\_upd} = t_k^{ts\_prev} - \left(VM_{ij}^{pp} \times Int_{CP}\right) \tag{7}$$

where $t_k^{ts\_upd}$ and $t_k^{ts\_prev}$ represent updated task size and previous task size, respectively. Due to the failure of a *PM* in the data center, its deployed VMs will also fail. If the failed VM is primary, the delay-sensitive task is migrated to one of the two backup VMs in the decreasing order of their reliability. In case, the backup VM selected for the task migration is already executing a delay-insensitive task, its task is suspended to make the backup VM available for the delay-sensitive task. A new idle backup VM is searched for the suspended task. Task migration subjected to some delay depends upon utilized memory, dirty memory page production rate, and transmission rate, etc. [11]. In E-FFTF, task migration delay $t_k^{del\_mig}$ is computed using pre-copy migration as

$$t_k^{del\_mig} = \frac{mem_{utl}}{rate_{trn}}\left(\frac{1 - R^{num_{itr}+1}}{1 - R}\right) \tag{8}$$

where mem_{utl} is the amount of utilized memory, $rate_{trn}$ is the rate of transmission, R is the ratio of dirty memory page production rate to transmission rate, and num_{itr} is the number of iterations required for migration. Considering the $t_k^{del\_mig}$, t_k^{ft} is modelled as

$$t_k^{ft} = t_k^{ft} + t_k^{del\_mig} \tag{9}$$

3 Experimental Evaluation

This Section evaluates E-FFTF through simulation experiments in terms of its fault tolerance capability, resource utilization, and price savings on both artificial and real workload. The experimental setup and simulation results are discussed as follows.

3.1 Simulation Settings

CloudSim [12] is used for the system performance evaluation with following simulation settings:

- Cloud environment consists of 10 data centers with 10000*PMs* each.
- *PM* failure is modelled on Gamma distribution [13] keeping $\alpha = 0.29$ and $\beta = 119.60$ like in [8].
- Eight types of *VMs* are considered with processing power as $250\,MIPS, 500\,MIPS, \ldots\ldots\ldots, 2000\,MIPS$.
- Time to activate a *PM* and deploy a *VM* is equal to $90\,and\,15\,seconds$, respectively [14].
- 50000, 100000, *and* 150000 independent delay-sensitive tasks are considered for execution.
- Google Workload Tracelog [15] is used to calculate the task size for real workload as

$$t_k^{ts} = (time_{finish} - time_{schedule}) \times u_{avg} \times p_{CPU} \tag{10}$$

  Where $time_{schedule}$, $time_{finish}$, and u_{avg} represent the schedule time, finish time, and average usage of CPU for the task, respectively. p_{CPU} (1500 *MIPS* [15]) is the CPU processing power.
- Task size for artificial workload is generated through uniform distribution across real Workload.
- For delay-sensitive tasks, deadlines are estimated corresponding to the *VMs*' average processing power. For delay-insensitive tasks, no deadline of completion is considered.
- Task arrival in artificial workload is modelled through Poisson distribution keeping average inter-arrival time as $2, 4,\ or\ 6\,seconds$. Task arrival in real workload is considered as per the Google Tracelog.
- Checkpoint interval equal to $60\,minutes, 90\,minutes, or 120\,minutes$ are considered to periodically save the execution states of running tasks.

To establish the fault tolerance capability, we evaluate its task deadline guarantee, and the resource utilization is realized through the group utilization index.

- *Task Deadline Guarantee (TDG)*: TDG is calculated as the percentage of delay-sensitive tasks successfully finished within their respective specified deadline.

$$TDG = \left(\frac{Number\ of\ delay - sensitive\ tasks\ meet\ their\ deadline}{Total\ number\ of\ delay - sensitive\ tasks} \times 100\right)\% \tag{11}$$

- *Group Utilization Index (GUI)*: GUI is calculated as the percentage of average time for which a replication group executes the assigned tasks (delay-sensitive + delay-insensitive) to the average time for which a replication group exists in the data center.

$$GUI = \left(\frac{Average\ execution\ time\ of\ CRG}{Average\ life\ time\ of\ CRG} \times 100 \right) \% \tag{12}$$

- *Average Execution Price (AEP)*: AEP measures the average price paid by the users for the task execution under a specific service class. For example, the AEP for Class A tasks is calculated as follows:

$$AEP_{ClassA} = \frac{\sum_{l=1}^{No.\ of\ Class\,A\ tasks} \{t_l^{ep}\}}{No.\ of\ Class\,A\ tasks} \tag{13}$$

3.2 *Results and Discussion*

Experiments are conducted to obtain the results in terms of TDG and GUI by altering one of the three parameters, viz. task count, task inter-arrival time, and checkpoint interval. Performance of E-FFTF is compared with FFTF in terms of fault tolerance (TDG) and resource utilization (GUI). Figures 1 and 2 compare the performance of E-FFTF and FFTF for the parameter TDG on artificial and real workload, respectively.

TDG of E-FFTF is comparatively higher than that of FFTF because service selection in E-FFTF is performed by the system based on the task slack time, whereas in FFTF, it is done by the user itself. Therefore, in FFTF a user may select a low fault tolerance service for price saving but end up in the task rejection due to low performance. Corresponding to the alteration in task count, TDG being above 95% proves its stability in varying situations. Along increasing checkpoint interval, TDG slightly decreases because at higher intervals, long rollbacks are required in the event of failures and the probability of missed deadlines increases. Along increasing task

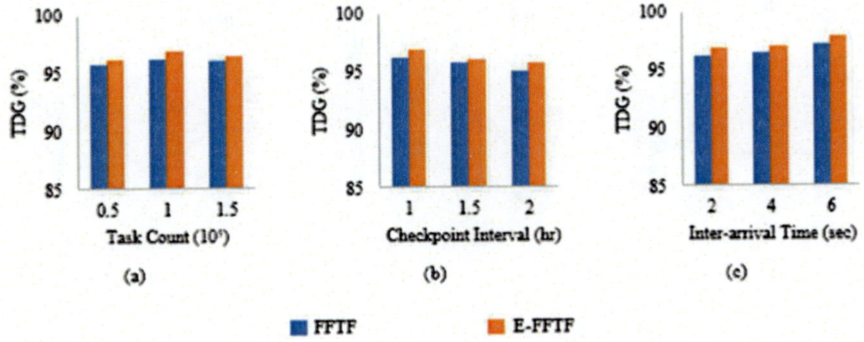

Fig. 1 TDG comparison on artificial workload w.r.t **a** task count, **b** checkpoint interval, and **c** inter-arrival

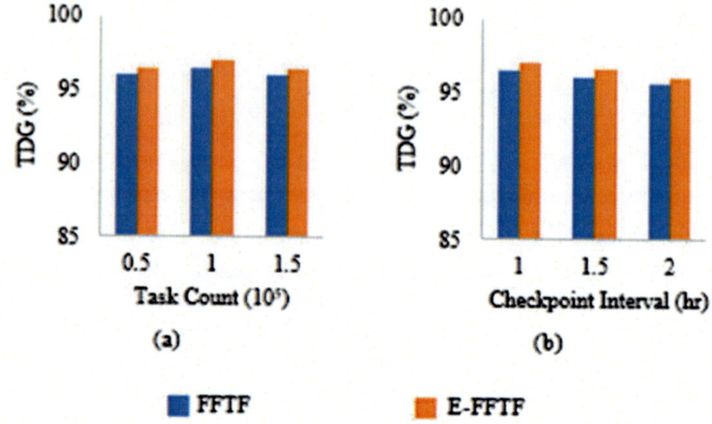

Fig. 2 TDG comparison on real workload w.r.t **a** task count and **b** checkpoint interval

inter-arrival time, TDG also increases because the system gets enough time to make the resources available.

Figures 3 and 4 compare the performance for GUI. In terms of resource utilization, both the frameworks achieve similar performance. As shown in figures, GUI increases along increasing checkpoint interval. It is explainable because at higher interval, long rollbacks are required which consequently increases the task execution time. On real workload, GUI shows little fluctuations which occur due to the large variation in the task size.

Figure 5 gives the AEP of E-FFTF for different service classes with respect to task count. AEP decreases along the decreasing reliability of service and hence justifies the price flexibility of the framework. Since, AEP is calculated considering the deadline of the task and reliability of the executing resources, therefore, in this

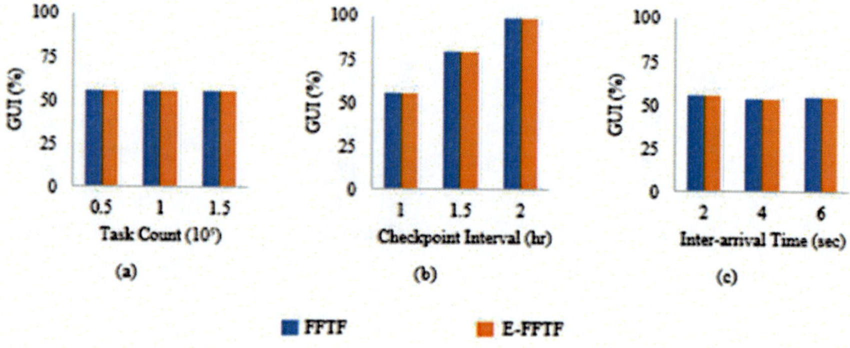

Fig. 3 GUI comparison on artificial workload w.r.t **a** task count, **b** checkpoint interval, and **c** inter-arrival time

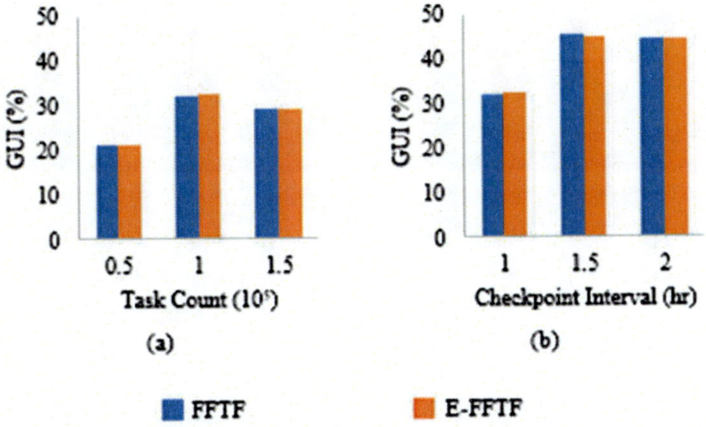

Fig. 4 GUI comparison on real workload w.r.t **a** task count and **b** checkpoint interval

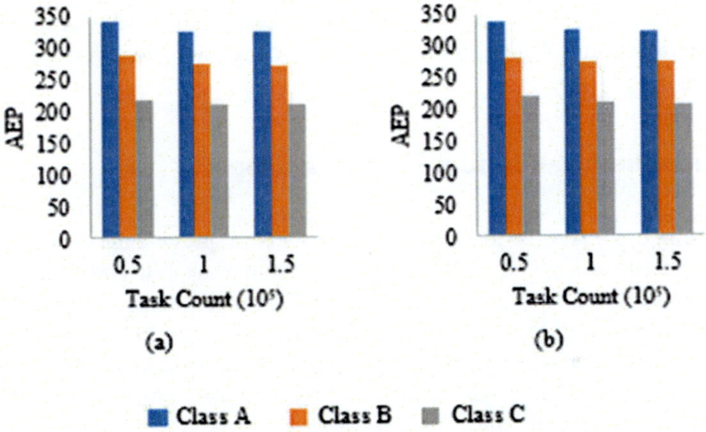

Fig. 5 AEP for different service classes on **a** artificial and **b** real workload

experiment, inter-arrival time and checkpoint interval are not varied as AEP is not dependent on them.

Figure 6 ascertains the AEP of E-FFTF and presents a comparative view with the non-flexible fault tolerance scenario. For comparison, the service reliability of 99.999% (reported for Amazon, Google and Microsoft) is extended to the users irrespective of their task deadline/slack time. Accordingly, the task execution price is computed for the high reliability service. In E-FFTF, comparatively low task execution price is obtained. Its prime reason is the saving in task execution price by selecting resources with reliability value directed by its slack time. Execution price of the tasks with enough slack time is reduced by selecting comparatively low reliability resources because their delayed execution may be accounted for its slack.

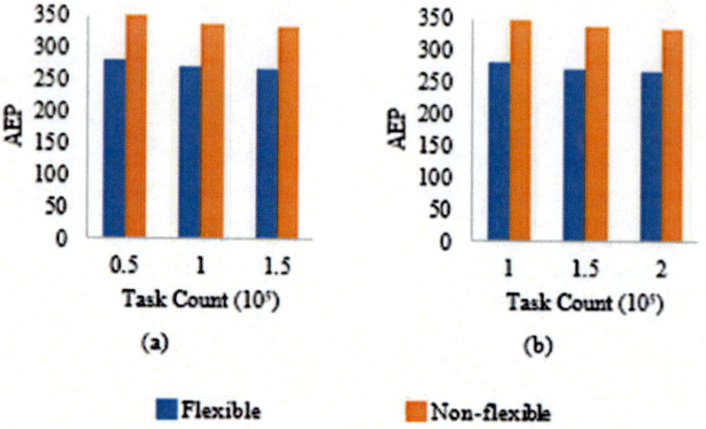

Fig. 6 AEP of flexible and non-flexible fault tolerance on **a** artificial and **b** real workload

Results shows that E-FFTF saves nearly 20% of the execution price as compared to the non-flexible fault tolerance without compromising the system performance.

4 Conclusions and Future Research Scope

In this research, FFTF is further extended to E-FFTF to provide the user transparent selection of service class and to establish its price saving capability while maintaining the performance. To prove the usefulness of E-FFTF, experiments are conducted over synthetic and real workload and obtained better results than FFTF. Task execution price saving is proved by comparing the total price invested in the flexible service selection through E-FFTF with the other well-established cloud service providers which intends to ensure high reliability service irrespective of the specified task deadline.

Prime outcomes from E-FFTF are: (a) Service selection is made transparent as class (A/B/C) is devised from the attributes such deadline and slack time. (b) Cloud user is benefitted by executing the task with softer deadline with low level of fault tolerance, which, in turn, levy low charges. (c) Service provider is benefitted by extending the services from low reliability resources to the user task with softer deadline. For the future research, we plan to apply the backup overloading approach within the replication groups along with executing delay-insensitive tasks to further enhance the resource utilization.

References

1. Fagg, G., Dongarra, J. (2000). FT-MPI: Fault tolerant MPI, supporting dynamic applications in a dynamic world. In *Proceedings of the Recent Advances in Parallel Virtual Machine and Message Passing Interface*. pp. 1–8
2. Abid, A., Khemakhem, M. T., Marzouk, S., Jemaa, M. B., Monteil, T., & Drira, K. (2014). Toward antifragile cloud computing infrastructures. *Procedia Comput. Sci., 32*, 850–855.
3. Yadav, N., & Goraya, M. S. (2018). Two-way ranking based service mapping in cloud environment. *Future Generation Computer Systems, 81*, 53–66.
4. Hasan, M., & Goraya, M. S. (2019). Flexible fault tolerance in cloud through replicated cooperative resource group. *Computer Communications, 145*, 176–192.
5. Goraya, M. S., & Kaur, L. (2013). Fault tolerance task execution through cooperative computing in grid. *Parallel Process. Lett., 23*, 1–20.
6. Malik, S., Huet, F., **2011.** Adaptive fault tolerance in real time cloud computing. In *Proceedings of the IEEE World Congress on Services*. pp. 280–287.
7. Varasteh, A., Tashtarian, F., Goudarzi, M. (2017). On reliability-aware server consolidation in cloud datacenters. In *Proceedings of the 16th International Symposium on Parallel and Distributed Computing*. pp. 95–101.
8. Birke, R., Giurgiu, I., Chen, L.Y., Wiesmann, D., Engbersen, T. (2014) Failure analysis of virtual and physical machines: Patterns, causes and characteristics. In *Proceedings of the Int. Conf. Dependable Syst. Networks*. pp. 1–12.
9. More Permutations and Combinations. https://www.cs.sfu.ca/~ggbaker/zju/math/perm-comb-more.html
10. Hartigan, A., Wong, M.A. (1979). A K-Means clustering algorithm. *J. R. Stat. Soc.*
11. Sun, G., Liao, D., Zhao, D., Xu, Z., & Yu, H. (2018). Live migration for multiple correlated virtual machines in cloud-based data centers. *IEEE Transactions on Services Computing, 11*, 279–291.
12. Calheiros, R. N., Ranjan, R., Beloglazov, A., De Rose, C. A. F., & Buyya, R. (2011). CloudSim: A toolkit for modeling and simulation of cloud computing environments and evaluation of resource provisioning algorithms. *Softw. - Pract. Exp., 41*, 23–50.
13. Schroeder, B., & Gibson, G. A. (2010). A large-scale study of failures in high-performance computing systems. *IEEE Trans. Dependable Secur. Comput., 7*, 337–350.
14. Hermenier, F., Lorca, X., Menaud, J., Muller, G., Lawall, J. (2009). Entropy: a consolidation manager for clusters. In *Proceedings of the 5th International Conference on Virtual Execution Environments*. pp. 41–50.
15. Wang, J., Bao, W., Zhu, X., Yang, T., & Xiang, Y. (2015). FESTAL: Fault-tolerant elastic scheduling algorithm for real-time tasks in virtualized cloud. *IEEE Transactions on Computers, 64*, 2545–2558.

Human Abnormal Activity Pattern Analysis in Diverse Background Surveillance Videos Using SVM and ResNet50 Model

S. Manjula and K. Lakshmi

Abstract Today, almost all the places are observed by surveillance cameras. The aim is to monitor the activities in and around, especially for abnormal activities. But it requires manual assistance to watch/monitor. However, manual inspection is a monotonous job, that reflects on information lost. It shows the importance of an automatic abnormal activity detection system. This is considered a difficult task because of object overlapping, light variation, clutter background, camera angle, presence of various activity, and posture variations. It manipulates the human actions in videos. Hence, it is challgening to recognize abnormal human actions. This work examines the concert of the classical *SVM* and *ResNet*50 model among four datasets: The '*SAIAZ*' (*Students Activities In Academic Zone*)-Corridor, '*SAIAZ*'-Open space, Classroom Violence from YouTube (cc), and Mixed Background Dataset (MBD). The '*SAIAZ*' was created by student volunteers of our Institution, and MBD is a collection of selected frames from various videos. Abnormal actions like slapping, punching, kicking, running, and fighting is commonly extant in these datasets. Here, MBD is assumed to adopt various real-world situations. The *SVM* achieved the classification accuracy of 85%, 92%, 60%, and 44% on *SAIAZ*-Corridor, *SAIAZ*-Open space, Classroom Violence, MBD respectively. The *ResNet*50 achieved significant improvements in all the given datasets.

Keywords Abnormal Activity detection · Surveillance videos · *SVM* Classifier · *ResNet*50 · Diverse background

---

S. Manjula (✉) · K. Lakshmi
Periyar Maniammai Institute of Science and Technology, Thanjavur, Tamil Nadu, India
e-mail: manjula_se@pmu.edu

K. Lakshmi
e-mail: lakshmi@pmu.edu

P. Nayak et al. (eds.), *IoT and Analytics for Sensor Networks*, Lecture Notes in Networks and Systems 244, https://doi.org/10.1007/978-981-16-2919-8_5

1 Introduction

With the incremental growth of the population, most of the public places are occupied by the people. It is forced to monitor the places and manage the crowd activity. It recommends increasing public safety and provide protection to the physical assets. Nowadays the cost and maintenance of the surveillance system are drastically reduced, and the rate of unwanted activities has increased that influence many places that are under surveillance. The surveillance system to be monitored continuously or read by security personnel, if necessary. However, analyzing the video manually is a tedious task because of the larger size of the videos and involving different kinds of activities. To avoid manual intervention, the system requires the automatic detection of abnormal activities [1]. So, video-based abnormal activity detection becomes an active research area [2]. The real-world actions are dissimilar in nature, thus could not define all the abnormal activities.

The aim of this work is to identify the abnormal activities like slapping, kicking, punching, running, and throwing a stone in the following datasets such as *'SAIAZ'* (*Student Activities In Academic Zone*)-Corridor,*'SAIAZ'-Open Space*, *Classroom Violence dataset*, and *Mixed Background Dataset (MBD)*. The *'SAIAZ'* dataset is taken from our university premises. The actions in 'SAIAZ' was performed by a group of 6 to 7 student volunteers. Classroom Violence is a video clip taken from YouTube (cc). The MBD is a collection of selected frames from different videos. The reason to have MBD is to assess the system performance in a heterogeneous background.

This work is intended to study the performance of the classical *SVM* model and *ResNet50* approach in the datasets *'SAIAZ'*, *Classroom Violence dataset,* and in *MBD*. The *SVM* classification model (Fig. 1) gets the input from the video sequences, then segmentation is performed by background subtraction after that statistical & Hu features are extracted [3], finally, classification is performed by *SVM* classifier.

Deep learning architectures are the most effective approaches for object detection and classification [4]. The CNN architectures have a deep structure, which learns

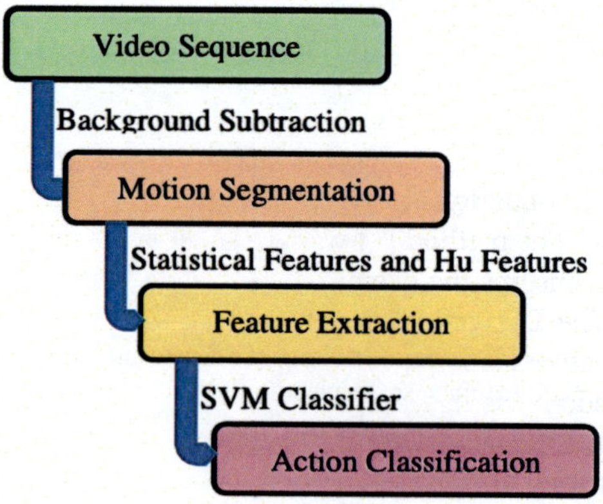

Fig. 1 SVM classification model

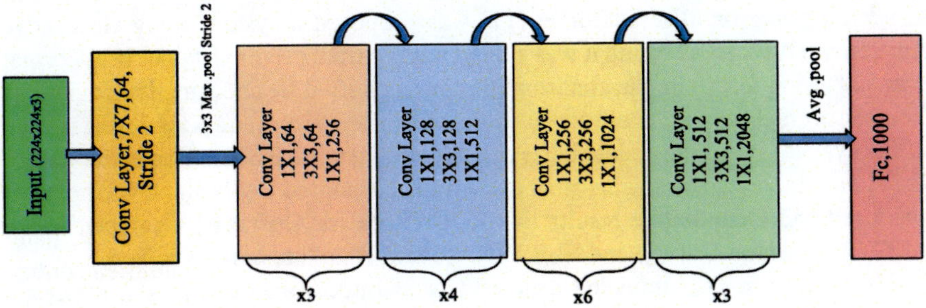

Fig. 2 Architecture of pre-trained *ResNet*50 model

high-level structures of input data. Deep learning algorithms achieved a great success in classifying actions [5]. The *ResNet50* deep learning model is used here. This model uses skip connections (Residual Connections) to avoid information loss during the training period. This network is a combination of 49 convolutional layers and 1 fully connected layer. Figure 2 shows the *ResNet50* architecture. This is a regular feed-forward neural network with a residual connection. This *SVM* and *ResNet50* aims to investigate the classification accuracy on various datasets. The paper is organized as follows. In Sect. 2, a review made on state-of-the-art abnormal action detection systems. Section 3 describes the working methodology. Section 4, Illustrate the experimental results of video sequences. Finally, Sect. 5 concludes and presents the future work.

2 Related Work

The amount of surveillance data is enormously growing in recent years. A well-organized and consistent video monitoring system is required [6] to handle this video data. In [7], a histogram of optical flow method was used as an image/frame descriptor. It is more reliable to notify the abnormal activities and the PETS benchmark dataset was utilized to validate the effectiveness of the system. In classification phase, a single and multi-view SVM classifier was used to classify the actions and it was found that running action as abnormal activity. The parameters such as ROC and AUC were used to evaluate the system performance. The system [8] considers the poor-quality, occluded images and maintains the smoothness of the event.

The method [9] was equipped with temporal dependencies between the actions; it ensures the event detection in videos, and it was designed to handle *null events*. The Dataset Hollywood and TRECVID Surveillance Event Detection (SED) was considered to evaluate the system performance. STIP and Fisher Vector features are adopted and a multiclass SVM classifier was encouraged to classify the actions. The author in [10] was designed a Fuzzy Semantic Petri Nets (FSPN) system. It aims to detect the concurrently occurring events, and gives a summary of the situation, of *what's going on.* The author in [11] was proposed a system using Generative Adversarial Nets (GAN), and it was skilled with normal events only not in abnormal events. During test time, the abnormalities are identified by checking the deviations from the normal actions. These can be assessed by frame-level and pixel-level fashion. In this system, a walking pedestrian is considered as a normal one and a person using a skateboard is treated as an abnormal activity. The standard UMN and UCSD dataset was used and Quantitative results like ROC (Receiver Operating Characteristics), AUC (Area Under Curve), and EER (Equal Error Rate) are employed here.

The author in [12] modeled a new FCN architecture to handle the abnormal regions. The kernels of this model are educated with the training video. This system is intended for speeder and precise system for anomaly detection in video data. The person walking in opposite direction is considered as an abnormal activity, walking pedestrian is a normal one.

In the system [13], a Temporal CNN Pattern was found the anomalous activity in a video data set. The walking pedestrian is considered as normal, movement of cars, bicycles, and skateboard is assumed as an abnormal activity. The UCSD and UMN standard dataset was used and the evaluation is performed on pixel level and frame level. This system's performance was evaluated by ROC, AUC, and EER.

The system [14] developed a new abnormal activity detection framework. It proposed Tracklets to formulae the salient points in each frame and the linear SVM was used to classify the actions in standard datasets like UMN, UCSD, and BEHAVE. The performance of the system was measured by ROC, AUC, and EER. It describes walking is a normal activity and people running away is an abnormal activity.

In this proposed Action Recognition with Image-Based CNN Features based system [15], where feature representation for videos is achieved by a hierarchical structure of CNN features such as CNN global features and *fully connected AlexNet features.* And it captures exact *AlexNet* features by using Caffe [16]. This proposed system achieves better performance on action recognition in the pioneered datasets such as KTH, UCF-11, and UCF-Sport action dataset. The system [17] 3D ResNet with Ranking Loss Function used to detect the abnormal activities in video data. This proposed 3D deep multiple instances learning with *ResNet* (MILR) model attains the best performance on the UCF-Crime dataset. ROC and AUC metrics are used to evaluate the system.

The author Hu [18] constructed a model to select the suspicious object region and the first-level features are extracted and fed into the deep network for deep feature extraction. The least-square SVM classification model is used for classification. The embedded chip Hi3531 was used and shows the high detection rate than the DLHS, DLPD, CTLD, Yolo-SD, and HOD-CCL methods. It detects the most abnormal behaviors and passes the alarming message in real-time surveillance.

3 Methodology

This proposed work is to study the performance of the *SVM* classification model and *ResNet50* model. And this is tested with *SAIAZ*, *Classroom Violence*, and *Mixed background Dataset (MBD)*. Each video dataset is converted to frames. The frames are labeled as a normal and abnormal activities. This presented system is trained with normal and abnormal frames and tested with test frames.

3.1 SVM Classifier Model

This work proposes to compare the activity identification using the classic *SVM* classifier and *ResNet50 model*. Well and profound *SVM* classifier is adapted and skilled to execute the linear and nonlinear classification [19, 20]. This *SVM* action classification system has the following stages 1. Obtaining video sequences, 2. Segmenting foreground objects by using background subtraction method, 3. Extracting features like statistical and Hu features by Hu invariant moments [3], 4. Classifying actions by *SVM* classifier. The part of this *SVM* work was published and available at: https://ssrn.com/abstract=3453542 [21].

3.2 ResNet50 *Model*

The convolutional neural networks (CNNs) are an optimal choice for analyzing and detecting an abnormal action in video sequence. The CNN architecture learns low-mid-high-level features from an image. The CNNs automatically extract features from the data. The CNN is a set of convolution, activation, pooling, and fully connected layers. The recent CNNs are deeper as compared to the shallow network. And the recent CNNs has complex connections among alternative layers like *ResNet* [17]. Deep learning algorithms attained prodigious success in detection and classification actions [22].

The Pre-trained *ResNet50* model is used to perform the classification task in this proposed work. The popular *Resnet50* model is a combination of 49 convolution layers and one fully connected layer [23]. The Fig. 2 shows the pre-trained *ResNet50* architecture. The *ResNet* was designed with ultra-deep network to handle the vanishing gradient problem that earlier model suffers [23]. The *Resnet* has a residual block is shown in Fig. 3.

The Data augmentation is accomplished to resize the image as 224 × 224 and the standard color augmentation is performed in the *ResNet50* model. As per the literature, the *LeNet* [24] is used for binary images of handwritten digits, whereas other pre-trained models are prepared to adopt to color images. After each convolution operation and before the activation function a batch normalization is performed,

Fig. 3 Residual block

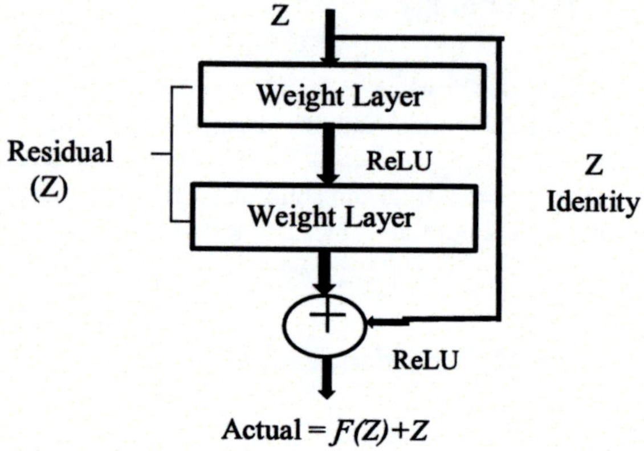

and the ReLU activation function is adopted here. The Mini-Batch and Stochastic Gradient Descent-based approaches are applied to train the data with a batch size 256 and along with the backpropagation mechanism. And this *ResNet50* model follows feed-forward neural network with *shortcut* connections. The *shortcut* connections of *ResNet50* model performs *identity* mapping and its output is added to the stacked layers.

Residual is an error [22] and this is the amount or number to change the prediction to meet out the actual value. The residual layer output is based on the outputs of $(l-1)$th layer. It is achieved by the previous layer of Z_{l-1} and $\mathcal{F}(Z_{l-1})$ is the output obtained by performing convolution operation with different sizes of filters. The predicted value 'Z' to be equal to the actual value. The residual function will compute and produce the residual of the model to match the predicted value with the actual value. By the mechanism of transfer learning, the same (pre-trained) deep learning approaches can be applied for different applications or with different types of data sets [23]. Therefore, ResNet50 pre-trained model is used to find the abnormal activities in the above-mentioned datasets.

4 Experimental Analysis

In this work, MATLAB 2018a is used to perform the experiments. The performance of *SVM* and *ResNet*50 model is compared with the datasets mentioned earlier. A set of normal and abnormal activity frames are obtained from the video data set. Four different types of video datasets were used here. *SAIAZ (Student Activities In Academic Zone)*—Corridor, *SAIAZ*-Open Space, *Classroom Violence, MBD* was utilized to test the performance of *SVM* Classifier and *ResNet50* deep learning model. The *SAIAZ* video dataset was created at our university premises by a group of six to seven student volunteers. It has two sets of video data: one is obtained from corridor and another one is from open-space. The *SAIAZ* dataset has 926 frames with the following activities like slapping, kicking, punching, hitting, and fighting. And it was obtained from a top angled single static camera with a homogeneous background and the backgroud illumination is uniform.

The *Classroom Violence* video dataset was taken from the YouTube (CC) and converted into frames. This video was taken from left, right, and front angled camera with different illumination condition. And this dataset consists 2433 frames with the activities like punching, kicking, slapping, pushing, and fighting. The MBD has 40 frames, it was formed by a set of selected frames from the different background videos. This MBD dataset has different kinds of background, illumination condition, variety of actions, involving a numerous number of persons with different posture.

All the video sequences are resized to 256×256 pixels. These four different kinds of data sets were tested with *SVM* classifier and *ResNet50* model. Figure 6 shows the *ResNet50* architecture obtained from MatLab. For simplicity, a specific region of *ResNet50* first layer is visualized. Figure 7, shows the low-level feature weights of first layer which was shown in the Fig. 6.

The dataset is classified as training set and test set. The training set has 100 frames, comprising of fifty normal activity frames and fifty abnormal activity frames, this was chosen by the presence of activity in the frame. The test set has 25 frames with 15 normal and 10 abnormal action frames. The systems performance is evaluated by the following performance metrics: True Positive Rate (TPR), True Negative Rate (TNR), False Positive Rate (FPR), False Negative Rate (FNR) (Table 2), Precision, Recall, F1 Score, and accuracy of action detection rate are used.

For *ResNet*50 model, the input image is resized as $224 \times 224 \times 3$ in data augmentation process and it is normalized with 'zerocenter' function. Here, the mean image is subtracted from an input image, to have a zero for mean and 1 for standard deviation. This ensures the smooth and faster optimization process during training period. Fc1000 is the output classification layer and it classifies the normal and abnormal actions. Figure 4a to 4h represents the sample normal and abnormal activity frames of four different kinds of datasets.

5 Result and Discussion

A portion of this work using *SVM* classification on *SAIAZ*-Corridor was published in [21], may refer for further details. This proposed work discusses about, how *SVM* and *ResNet*50 works on different types of dataset. In the *ResNet50*, the residual block is a basic structure of network. In the training period, a certain residual block is enabled, and the input streams are passed through the identity shortcut and the weight layers, otherwise the input flows only through the identity shortcut. Figure 7, shows the low-level feature weights (edges, curves) of the first layer. During training time, the layer has a "existence possibility" and it is randomly dropped. At the time of testing, all blocks are kept active and re-calibrated according to its existence possibility. The camera angle, illumination, background, and posture-based features of images

Fig. 4 **a, b**—*SAIAZ*-corridor: normal and abnormal action frames; **c, d**—*SAIAZ*-open space: normal and abnormal action frames; **e, f**—classroom -violence—normal and abnormal action frames; **g, h**—mixed background—normal and abnormal action frames

are used for the classification task. The classification accuracy is presented in the Table 1.

In the dataset *SAIAZ*-Corridor and *SAIAZ*-Open space, the proposed model *SVM* and *ResNet*50 works well when compared to the remaining two datasets. The observation presented from the above video dataset is taken from the top angled static camera and there is no angle transformation in it. It is also observed that the illumination is uniformly distributed and the background is identical all over the video. The accuracy of the *SVM* on *SAIAZ*-Corridor is relatively lower than the *SAIAZ*-Openspace, Why, because the training set of *SAIAZ*-open space was clearly defined. The accuracy was brought down in *SAIAZ* Corridor due to ambiguity in a selected training set. Where the deep learning model gets better results on it. Looking into the Classroom-violence dataset, the detection rate of *SVM* is reduced because the dataset is taken from a different angle camera. Here, the illumination is diverse and more objects are involved; these are few notable reasons why it brings down the detection rate. Even in the uncertainty of datasets, the *ResNet50* works well. In the above three cases, each training dataset and testing dataset is taken from the same video. But in the case of MBD, the system is trained with any one of the above three dataset, but the test case is from MBD, hence the detection rate of the system is reduced. In

Table 1 Classification accuracy obtained on various dataset

Dataset	*SAIAZ*-Corridor (SC)		*SAIAZ*—OpenSpace (SO)		Classroom Violence (CV)		Mixed Background (MBD)	
	SVM	ResNet50	SVM	ResNet50	SVM	ResNet50	SVM	ResNet50
% of Accuracy	85	96	92	96	60	92	44	68

Table 2 Performance metrics

Dataset	*SAIAZ*- Corridor (SC)		*SAIAZ*—OpenSpace (SO)		Classroom-Violence (CV)		Mixed Background Dataset (MBD)	
Metric/Model	SVM	ResNet50	SVM	ResNet50	SVM	ResNet50	SVM	ResNet50
TPR	1	1	0.9	1	0.6	0.9	0.75	0.8
FNR	0	0	0.1	0	0.4	0.1	0.25	0.2
TNR	0.73	0.93	0.93	0.93	0.6	0.93	0.16	0.6
FPR	0.33	0.07	0.06	0.07	0.4	0.06	0.84	0.4

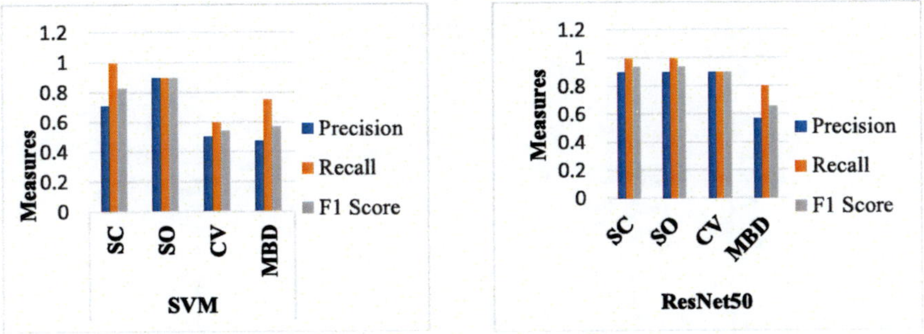

Fig. 5 **a** Performance Measures on SVM. **b** Performance Measure on ResNet50

the fourth case as well, deeper layers have learned the features and have produced positive results than the SVM.

In Table 2, the performance of the proposed model is displayed. The True Positive Rate (TPR), False Negative Rate (FNR), True Negative Rate (TNR), and False Positive Rate (FPR) is obtained. In Fig. 5a and 5b, the performance metrics precision, recall, and F1 Score are exhibited. Figure 8 shows some of the mixed background resultant video frames. As per the results, comparatively *ResNet50* shows the better performance on all type of datasets (Fig. 5a and 5b).

$$TPR = TP/(TP + FN) \tag{1}$$

$$FNR = FN/(TP + FN) \tag{2}$$

$$TNR = TN/(TN + FP) \tag{3}$$

$$FPR = FP/(FP + TN) \tag{4}$$

$$\text{Precision} = TP/(TP + FP) \tag{5}$$

$$\text{Recall} = TP/(TP + FN) \tag{6}$$

$$F1\ \text{Score} = 2 \times \frac{\text{Precision} \times \text{Recall}}{\text{Precision} + \text{Recall}} \tag{7}$$

$$\text{Accuracy} = TP + TN/N \tag{8}$$

TP—True Positive; FN—Fales Negative; TN—True Negative; FP—False Positive; N—Total number of frames.

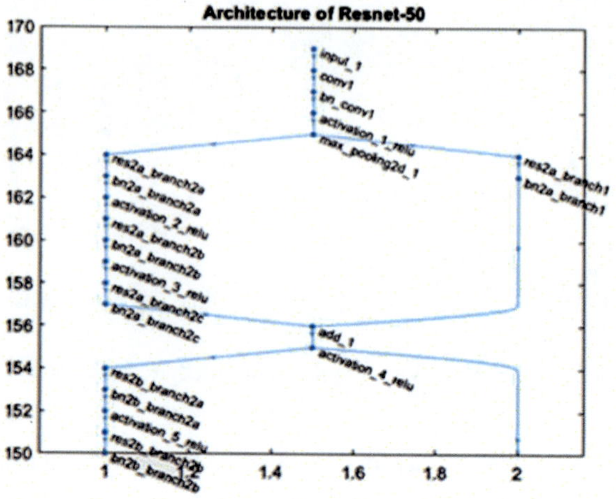

Fig. 6 First Layer of ResNet50 on Visualization of region 50–170 obtained through MATLAB

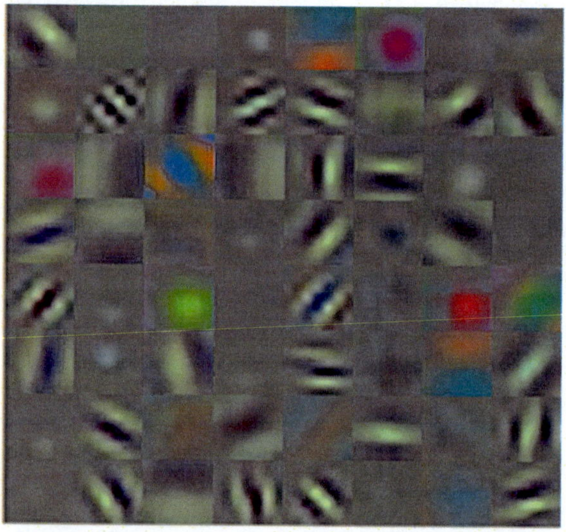

Fig. 7 Visualization of low-level feature weights (edges, curves) of first layer of region 150–170 obtained through MATLAB

6 Conclusion

This system evaluates the classification accuracy in classical *SVM* classifier and *ResNet50* model. The *SVM* gives better classification accuracy on *SAIAZ*, where

$$\delta V_{\rm hd} = -g_s^{-3/2} \frac{3^4 \lambda W_0^4}{\mathcal{V}^4} \Pi_i t^i \,, \tag{7.4}$$

where λ is an undetermined combinatorial number and t^i are the 2-cycle volume moduli. Π_i are topological quantities defined in terms of the (1, 1) forms $\hat{D}_i$ as

$$\Pi_i = \int_{\mathcal{X}} c_2 \wedge \hat{D}_i \,, \tag{7.5}$$

with c_2 the second Chern class of $\mathcal{X}$. Although these effects enter at higher F-term order, they can become important and comparable to string loop corrections.

- *Non-perturbative corrections*
  These are corrections to the superpotential [9]

$$W \to W_0 + \sum_i A_i \, \mathrm{e}^{-a_i T_i} \,, \tag{7.6}$$

  related to the existence of E3-brane instantons ($a_i = 2\pi$) or gaugino condensates on D7-branes ($a_i = 2\pi/N$, where N is the rank of the condensing gauge group). Similar non-perturbative corrections to the Kähler potential are subleading when compared to perturbative terms, which can arise from expansions in α' or g_s, and will thus be unimportant for our discussion here.

Together these corrections yield a scalar potential:

$$V = \delta V_{\alpha'} + \delta V_{g_s} + \delta V_{\rm hd} + \delta V_{\rm np} \tag{7.7}$$

where $\delta V_{\rm hd}$ is given by Eq. (7.4) and (setting $\mathrm{e}^{K_0} = 1$)

$$\delta V_{\alpha'} = \frac{3\hat{\xi} W_0^2}{4\mathcal{V}^3} \tag{7.8}$$

$$\delta V_{g_s} = \frac{W_0^2}{\mathcal{V}^2} \sum_i \left(g_s^2 C_i^2 \, K_{ii}^{\rm tree} - 2\frac{\tilde{C}_i}{\mathcal{V} t_i^{\cap}} \right) \tag{7.9}$$

$$\delta V_{\rm np} = \sum_{i,j} K^{ij}_{\rm tree} a_i a_j A_i A_j \frac{\mathrm{e}^{-a_i T_i - a_j \bar{T}_j}}{\mathcal{V}^2} + \frac{2 W_0}{\mathcal{V}^2} \sum_i A_i a_i \tau_i \left(\mathrm{e}^{-a_i T_i} + \mathrm{e}^{-a_i \bar{T}_i} \right) \tag{7.10}$$

$K_{ij}^{\rm tree}$ is the tree-level contribution to the Kähler metric, and $K^{ij}_{\rm tree}$ its inverse. Notice how the Kaluza-Klein contribution to Eq. (7.9) enters at second order thanks to the 'extended no-scale structure' [7]. All these corrections are under control when the overall volume $\mathcal{V}$ is large. In the regime where all 2-cycles scale as $t \sim \sqrt{\tau} \sim \mathcal{V}^{1/3}$, we have the scaling

$$\delta V_{\alpha'} \sim \frac{W_0^2}{\mathcal{V}^3} \,, \quad \delta V_{g_s} \sim \frac{W_0^2}{\mathcal{V}^{10/3}} \,, \quad \delta V_{\rm np} \sim \frac{\mathcal{V}^{4/3} \mathrm{e}^{-2a\tau} + W_0 \mathcal{V}^{2/3} \mathrm{e}^{-a\tau}}{\mathcal{V}^2} \,, \quad \delta V_{\rm hd} \sim \frac{W_0^4}{\mathcal{V}^{11/3}} \,, \tag{7.11}$$

where we have taken $K_{ij}^{\text{tree}} \sim 1/\mathcal{V}^{4/3}$. To generate stable vacua, one has to find a balance between different terms in the potential. For example, in KKLT models [10], W_0 is tuned to exponentially small values, $W_0 \sim (a\tau)\mathrm{e}^{-a\tau} \ll 1$, so that the two contributions to δV_{np} are comparable in size. This typically yields a supersymmetric AdS vacuum whose depth is parametrised by $-W_0^2/\mathcal{V}^2$. Upon uplift the same scale controls the height of the barrier separating the vacuum from the decompactification limit [10]. A notable exception to this rule is the racetrack setup which we describe in Sect. 8.2.1 [11, 12], where the scale of the vacuum can be made arbitrarily small thanks to two instanton contributions that are aligned relative to one another.

Another possible approach is to balance perturbative against non-perturbative corrections. Generically we expect the latter to be suppressed, except in the presence of small cycles, as this raises the size of the instanton correction. This is precisely what happens in the LVS scenario [13–15], where $\delta V_{\alpha'} \sim \frac{W_0^2}{\mathcal{V}^3} \sim \delta V_{\text{np}}$. This sets the scale of the potential, controlling both the depth of the non-supersymmetric AdS vacuum and the height of the barrier to infinity which develops after uplifting.

Before moving onto the model building aspects, we would like to review in a bit more detail each of these corrections.

7.2 Non-renormalisation Theorems

A first consideration might be to ask why are there no effects at $\mathcal{O}(\alpha')$ or $\mathcal{O}(\alpha'^2)$ in the type IIB supergravity theory. In turns out that the linear and quadratic order corrections in α' are proportional to the Ricci tensor of the internal manifold. For a Calabi-Yau compactification, the internal geometry is Ricci-flat and this corrections vanish as long as the internal manifold remains Calabi-Yau. This result is due to the sigma model calculations of [16], to four-loops in the worldsheet action, and the non-renormalisation theorems of [17]. An explicit description of the non-renormalisation theorems for type IIB supergravity can be found in [18].

Below we concern ourselves with a simpler but qualitatively identical question. We consider the four-dimensional effective theory for the heterotic superstring as described in Sect. 6.2.3. The supergravity approximation is controlled by the string coupling, which maps to the vev of the real part of the axio-dilaton field $\mathrm{e}^{\phi} = Re(S)$, and the α' parameter, which maps to the vev of an overall Kähler volume modulus $\mathcal{V}$. The low-energy dependence of the compactified four-dimensional effective theory on the internal geometry is encapsulated by three functions: the Kähler potential $K(S, T, U)$, the superpotential, which for heterotic supergravity has a no-scale structure in the axio-dilaton and Kähler directions $\Phi := \{S, T_\alpha\}$; and the gauge-kinetic coupling matrix $\mathcal{M}_{\kappa\lambda}$ depending only on the complex structure sector by virtue of being its kinetic metric.[1] The scalar potential, in the absence of D-terms, is given by the Kähler potential and the superpotential only, $V = V(K, W)$.

[1] This is guaranteed to hold as long as the moduli space is a direct product between the complex structure and the quaternionic sector.

We will argue that the superpotential is protected by a number of shift symmetries in the moduli fields and becomes exact at any order in the perturbative expansion, obtaining only non-perturbative corrections. In the case of heterotic supergravity, the discussion is quite simple. We note that the axionic partners to the dilaton and the overall volume modulus benefit from an unbroken shift symmetry meaning that the supergravity action is invariant under transformations of the form $\Phi \to \Phi + i\lambda$ where $\lambda \in \mathbb{R}$. Supersymmetry implies that the superpotential W and the matrix $\mathcal{M}_{\kappa\lambda}$ are holomorphic functions of the moduli fields. The axion shift invariance then forbids W from depending on the imaginary part of Φ and thus, by virtue of these functions being holomorphic, it cannot depend on the real parts either at any order in the perturbative expansion. Then, any perturbative corrections to the scalar potential of the tree-level theory must arise from the Kähler potential itself,[2] and these enter the theory at α'^3 as we will see in the next section. A similar argument can be made in more generality to show that the gauge-kinetic coupling matrix only receives corrections at first order in the loop expansion from Kähler moduli dependent contributions, in agreement with the calculations of [19].

The key point in supporting the previous arguments, and also in the cases of type II theories, is the survival of the axionic shift invariance in the presence of quantum corrections. This axionic symmetry can be understood as a subgroup of the larger accidental symmetry $SL(2,\mathbb{R})$ of the full string theory. In particular, the ten-dimensional type IIB supergravity action in Eq. (2.2) can be made explicitly invariant under $SL(2,\mathbb{R})$ by exchanging the Kalb-Ramond and RR three-form field strength by their gauge invariant counterpart in $\hat{G}_3 := \hat{F}_3 - iS\hat{H}_3$. With this choice of variables, we lose an explicit description of the Kähler and complex structure sectors in terms of chiral multiplets, in contrast to Eq. (2.43), but we can see the symmetry group for the axio-dilaton explicitly. Indeed, the contents of the ten-dimensional theory in this form are given by $\hat{g}_{MN}$, $\hat{Q} := i\hat{S} = \hat{C}_0 + ie^{-\hat{\phi}}$, $\hat{G}_3$ and the imaginary self-dual five-form $\hat{F}_5$. Under an $SL(2,\mathbb{R})$ transformation [20], the metric and the imaginary self-dual five-form remain invariant while

$$\hat{Q} \to \frac{a\hat{Q}+b}{c\hat{Q}+d}, \qquad \hat{G}_3 \to \frac{\hat{G}_3}{c\hat{\tau}+d}, \tag{7.12}$$

for $ad - bc = 1$ and $a, b, c, d \in \mathbb{R}$. The axion shift symmetry corresponds to a subset $SL(2,\mathbb{Z}) \subset SL(2,\mathbb{R})$ with $c = 0$ and $a = d = 1$ for which $\hat{Q} \to \hat{Q} + b$. Critically, $SL(2,\mathbb{Z})$ is expected to be an invariance of string theory even in presence of non-perturbative effects and the perturbative non-renormalisation of the type IIB supergravity will be preserved. We remark that this does not prevent the superpotential from acquiring non-perturbative corrections, cf. Sect. 7.6, but these will be exponentially suppressed in a large volume/weak coupling limit.

[2] As long as supersymmetry is not explicitly broken.

7.3 α'^3 Corrections

We begin by discussing the corrections to the Kähler potential to $\mathcal{O}(\alpha'^3)$ as calculated in [21, 22]. The goal of this section would be to analyse the effects of the higher order terms appearing in the M-theory effective action on the scalar potential of type IIB. However, only a few results regarding these terms are known and the best approach to obtain the leading order corrections to the type IIB action when compactifying on a CY is to use mirror symmetry and the c-map of [23].

In the seminal work [22], the authors provide an exact result for the leading corrections to the Kähler potential for the quintic and its mirror. In particular, the Kähler prepotential for the Kähler deformations with the leading order perturbative contributions is given by

$$f(t) = -\frac{1}{6}\mathcal{K}_{ABC}\frac{t^{\hat{A}}t^{\hat{B}}t^{\hat{C}}}{t^0} - (t^0)^2\frac{\hat{\xi}}{4}\,, \tag{7.13}$$

where $t^{\hat{A}} = \{t^0, t^A\}$ are the adapted coordinates described in Sect. 2.1. Using the definition of the Kähler potential in terms of its prepotential we find Eq. (7.2). Although the precise calculations are quite involved, we can provide a minimal qualitative derivation of the formula above.

As our target manifold $\mathcal{M}$, we take the quintic $\mathbb{P}_4(5)$ which has two neat properties. First, its mirror, which we will take as the base manifold and denote by $\mathcal{W}$, is known thanks to the construction of [24]. Second, the Betti numbers for this threefold are $b_{1,1} = 1$ and $b_{2,1} = 101$. This allows for simple calculations in the Kähler sector of the target manifold. The mirror of the quintic will have $b_{1,1} = 101$ and $b_{2,1} = 1$, which will simplify the calculations in the complex structure sector of the base manifold. We will denote the unique Kähler (complex structure) coordinate of the target (base) manifold by ψ.

We consider a one-parameter family of quintic hypersurfaces as a set in $\mathbb{CP}^4$ given by the roots of the polynomial

$$p = \sum_{k=1}^{5} x_k^5 - 5\psi\prod_{k=1}^{5} x_k\,, \tag{7.14}$$

which are invariant under the symmetry group generated by

$$g_0 = (1,0,0,0,4)\,,\quad g_1 = (0,1,0,0,4)\,,\quad g_2 = (0,0,1,0,4)\,,$$
$$g_3 = (0,0,0,1,4)\,, \tag{7.15}$$

where, for example, g_3 represents the $\mathbb{Z}_5$ action

$$(x_1, x_2, x_3, x_4, x_5) \to (x_1, x_2, x_3, \alpha x_4, \alpha^4 x_5)\,, \tag{7.16}$$

on the set of coordinates x_k on the quintic and $\alpha = e^{2i\pi/5}$. The geometry of the complex structure in $\mathcal{W}$ is determined by the coordinate ψ. It is most natural to consider ψ^5 as the coordinate of the complex structure sector. This can be seen by noting that a scaling of ψ to $\alpha\psi$ is equivalent to the coordinate transformation[3]

$$(x_1, x_2, x_3, x_4, x_5) \to (\alpha^{-1}x_1, x_2, x_3, x_4, x_5) \,, \tag{7.17}$$

which leads to the sample complex structure for the quintic in Eq. (7.14). Furthermore, the complex structure geometry of $\mathcal{W}$ is singular for some values of ψ. Since we are interested in describing a Calabi-Yau we will have to smooth out the metric around the singularities. It is, therefore, instructive to describe the geometry in a neighbourhood of these singular points. To do so we note that on can identify the set of singular points of $\mathcal{W}$ with the points where the quintic fails to be transverse [22]. This is the case when the equations

$$\partial_{x_k} p = 0 \,, \tag{7.18}$$

are satisfied simultaneously and we find

$$\prod_{k=1}^{5} x_k^5 = \psi^5 \prod_{k=1}^{5} x_k^5 \,, \tag{7.19}$$

which will only be satisfied for $\psi^5 = 1$ and this in turn implies $x_k = \alpha^{n_k}$ with $\sum n_k = 0$. All these points can be identified through the generators of Eq. (7.15), and $\mathcal{W}$ at $\psi = 1$ only has one singular point. In [25], it was shown that this singularity corresponds to a conifold for $\mathcal{W}$, where the singular point of the conifold is identified with the $\psi = 1$ point of $\mathcal{W}$. Locally, the geometry is given by a cone with base $S^2 \times S^3$ where, at the “tip” of the cone, the S^3 shrinks to zero volume with radius $r_{S^3} \sim \mathcal{O}(\sqrt{\psi - 1})$.

Now we can define the complex structure sector of the base manifold. We choose a symplectic basis $\{A^1, A^2, B_1, B_2\}$ for $H^3(\mathcal{W})$ such that the only non-vanishing intersections are $A^{\hat{K}} \cap B_{\hat{L}} = \delta^{\hat{K}}_{\hat{L}}$, and the dual basis $(\alpha_{\hat{K}}, \beta^{\hat{K}})$ such that

$$\int_{A^{\hat{L}}} \alpha_{\hat{K}} = \delta^{\hat{L}}_{\hat{K}} \,, \qquad \int_{B_{\hat{L}}} \beta^{\hat{K}} = \delta^{\hat{K}}_{\hat{L}} \,. \tag{7.20}$$

Finally, the holomorphic three-forms can be expanded as

$$\Omega = \mathcal{Z}^{\hat{K}} \alpha_{\hat{K}} - \mathcal{F}_{\hat{K}} \beta^{\hat{K}} \,. \tag{7.21}$$

[3] This is equivalent to the coordinate transformations $(x_1, \alpha^{-1}x_2, x_3, x_4, x_5) = \cdots = (x_1, x_2, x_3, \alpha^{-1}x_4, x_5)$ by the actions of the generators Eq. (7.15). This freedom is due to the polynomial p defining a covering space of $\mathcal{W}$, rather than $\mathcal{W}$ itself.

We remark that we have 4 periods $(\mathcal{Z}^1, \mathcal{Z}^2, \mathcal{F}_1, \mathcal{F}_2)$ and only one parameter ψ. In the usual way discussed in Sect. 2.32, we can fix half of the periods in terms of the other half and we choose $\mathcal{F}_{\hat{K}} := \mathcal{F}_{\hat{K}}(\mathcal{Z}^{\hat{L}})$. Furthermore, Eq. (2.32) implies that the last free period can be fixed by absorbing the gauge freedom of complex rescaling of the holomorphic three-form.

We now discuss the monodromy structure around the singularity. Following the conventions of [22], we take A^2 to be the cycle related to the vanishing S^3 at $\psi = 1$. By our choice of symplectic basis, the cycles (A^1, B_1) do not intersect with the S^3 and can be taken to lie outside the neighbourhood of the singularity. The last cycle B_2 intersects A_2 at a point, in the neighbourhood of the conifold singularity. Under transport around $\psi = 1$, B_2 can pick a multiple of A_2 while the other cycles are unambiguously defined. Then,

$$\Pi := \begin{pmatrix} \mathcal{Z}^1 \\ \mathcal{Z}^2 \\ \mathcal{F}_1 \\ \mathcal{F}_2 \end{pmatrix} \rightarrow \begin{pmatrix} 1 & 0 & 0 & 0 \\ 0 & 1 & 0 & 0 \\ 0 & 0 & 1 & 0 \\ 0 & n & 0 & 1 \end{pmatrix} \begin{pmatrix} \mathcal{Z}^1 \\ \mathcal{Z}^2 \\ \mathcal{F}_1 \\ \mathcal{F}_2 \end{pmatrix} , \tag{7.22}$$

where we have defined the period vector Π. The Kähler potential is given by

$$\mathrm{e}^{-K} = \bar{\mathcal{Z}}^{\hat{K}} \mathcal{F}_{\hat{K}} - \mathcal{Z}^{\hat{K}} \bar{\mathcal{F}}_{\hat{K}} \tag{7.23}$$

and by making use of the monodromy structure, the asymptotic form of the Kähler metric as $\psi \rightarrow \infty$ was calculated in [22] to be

$$g_{\psi\bar{\psi}} = \frac{3}{4\,|\psi|^2\,(\ln|5\psi|)^2}\left(1 - \frac{48\zeta(3)}{25(\ln|5\psi|)^3} + \cdots\right) , \tag{7.24}$$

where the ellipsis denotes further subleading terms in $1/\ln|\psi|$. The first term takes the usual inverse quadratic form upon the identification

$$t \sim \frac{\sqrt{5}}{2\pi}\ln(5\psi) , \qquad g_{\psi\bar{\psi}} \sim \frac{3}{4v^2} dt \wedge \star dt , \tag{7.25}$$

where we have chosen the evocative notation of $v := \mathrm{Re}(t)$ as the ψ in this regime will match the large volume limit on the dual side. Turning to the target quintic, the prepotential for the Kähler sector prior to any corrections is given by the first term in Eq. (7.13), the results can be matched to the ones obtained in the dual side for the asymptotic metric in Eq. (7.25). The full details of this matching can be found in [22] and references therein. The second term in Eq. (7.24) will be related with the Kähler corrections that we are looking for.

Turning once again to the target quintic, the non-renormalisation theorem of [17] imply that the are no σ-model α' corrections to the prepotential couplings

$$\partial^3 f(t)/\partial t^a \partial t^b \partial t^c \ .$$

Furthermore, the prepotential being a homogeneous function of degree two in the t^a coordinates is determined by the third derivative up to a quadratic term. This translates into the Kähler potential being fixed by Candelas et al. [22], Grimm [26]

$$\mathrm{e}^{-K} = -2t^{\hat{A}} \mathrm{Re}\,(\mathcal{N}_{\hat{A}\hat{B}}) t^{\hat{B}} \ , \tag{7.26}$$

where the matrix $\mathcal{N}_{\hat{A}\hat{B}}$ is the gauge-coupling matrix given in Eq. (2.24), which is a matrix constructed from the second derivatives of the prepotential $f(t)$. We note that the imaginary part of $\mathcal{N}_{\hat{A}\hat{B}}$ drops out of the formula above. This can be understood by virtue of the complex rescaling invariance of Ω, an imaginary shift in the matrix $\mathcal{N}_{\hat{A}\hat{B}}$ can be reabsorbed into a redefinition of the t^0 coordinate. A shift to the real part, however, will map to a change in the volume of the quintic $\mathcal{V}(\mathbb{P}_4(5)) \sim \mathrm{Re}\,(t)^3 \sim v^3$ which, based on the mapping in Eq. (7.25), will affect the metric of the Kähler sector and its complex structure dual. This will induce a correction to the prepotential function $f(t)$.

The first non-trivial contribution is unique at $\mathcal{O}(\alpha'^3)$ and corresponds to the combination

$$\mathbb{S} = \frac{1}{12(2\pi)^3} R_{IJ}{}^{KL} R_{KL}{}^{MN} R_{MN}{}^{IJ} - 2R_I{}^K{}_J{}^L R_K{}^M{}_L{}^N R_M{}^I{}_N{}^J \ , \tag{7.27}$$

which for a six-dimensional manifold corresponds to its Euler density,

$$\int_{\mathcal{X}} d^6x \sqrt{g}\, \mathbb{S} = \chi \ . \tag{7.28}$$

This type of contribution leads to corrections to the gauge kinetic coupling $\delta\mathcal{N}_{\hat{A}\hat{B}}$ which translate into a constant shift to the prepotential of the desired form upon dimensional reduction to the four-dimensional theory [21], i.e.

$$\delta f(t) = -\mathrm{e}^{-1/2\phi_0} \frac{\hat{\xi}}{2} \ . \tag{7.29}$$

This expression matches Eq. (7.13) up to an overall constant scale ϕ_0, which can be absorbed in the definition of t^0.

Although the calculations of [21, 22] are done on a pair of quintics, the results are independent of this choice as long as mirror symmetry is a true symmetry of string theory. The choice of the quintic allows for explicit calculations and it is argued in [21, 22] that the form of the corrections to the prepotential will generalise to other smooth Calabi-Yaus.

7.4 One-Loop Corrections

In this section we concentrate on corrections due to the exchange of KK or winding modes. In [27], the authors provide the one-loop string corrections to type IIB compactified on toroidal orientifolds by evaluating string scattering amplitudes. In [28], the authors study the consistency of the perturbative expansion of the string effective action in the presence of D-branes and O-planes. In particular, the authors make an educated guess for the generalisation of the toroidal corrections to smooth Calabi-Yaus and are able to obtain the characteristic scaling of such corrections and show that these are subleading to the α'-corrections discussed in the previous section. This also served as a consistency check for the LVS scenario of [29] and which we discuss in Sect. 7.10. Below, we provide a short overview of these arguments.

We begin by summarising the results of [27] for the toroidal orientifold. There, the Kähler potential takes the form[4]

$$
\begin{aligned}
K = & -2\ln(S+\bar{S}) - 2\ln(\mathcal{V}) + K_{cs}(U,\bar{U}) - \frac{\hat{\xi}}{2\mathcal{V}} \\
& + \sum_i g_s \frac{C_i(U,\bar{U})}{4\tau_i} + \sum_i \frac{\tilde{C}_i(U,\bar{U})}{4\tau_i\tau_j} \,.
\end{aligned}
\tag{7.30}
$$

The first line contains the tree-level contributions to the Kähler potential and the α'-corrections previously discussed. The second line are the corrections to the Kähler potential in the presence of localised sources (Fig. 7.1). In the following we exemplify the mechanisms with D-branes, but the statements would hold true for O-planes as well.

The first term is due to parallel sources (D7/D3-branes or O7/O3-planes) exchanging KK modes between them. The D7-brane is wrapped around some internal four-cycle τ_i where the point-like component of the D3-brane also lies. The two branes can then interchange closed strings along some two-cycle t_i.

The second term is due to the intersection of D7-branes (O7-planes) wrapped around four-cycles $\tau_i \neq \tau_j$. The four-cycles will intersect in some two-cycle t_i. If, locally, this t_i contains non-trivial one-cycles the winding modes can wound around them. It could seem that these modes would not appear in the CY generalisation. However, we note that from the point of view of the open string, these mechanism is based on Dirichlet strings that have their endpoints glued to each D7-brane. Therefore, the condition on the existence of these one-cycles is not a condition on the global geometry of the Calabi-Yau, but rather a condition on the topology of specific cycles within cycles. If the particular model to be considered contains no intersecting D7-branes, these corrections will vanish.

In [28], it is then shown that the scaling in smooth Calabi-Yaus is given in terms of the internal volume $\mathcal{V}$ rather than particular four-cycles. In particular, these corrections take the form given in Eq. (7.3)

[4] Where we have adapted the notation to be consistent with the rest of this chapter.

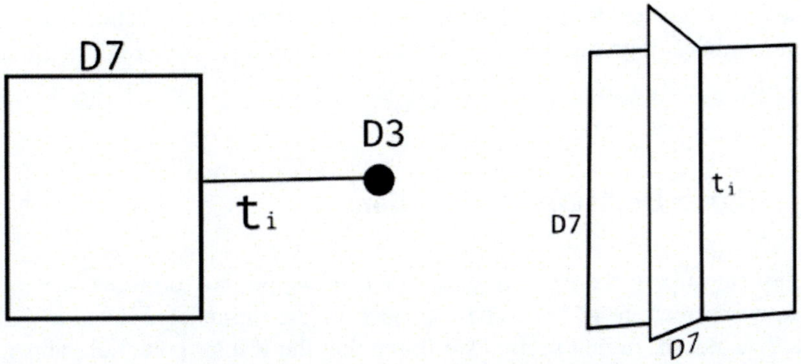

Fig. 7.1 Left: pictorial representation of the KK mode exchange mechanism for a coincident D7/D3-branes which can be separated by some distance parametrised by a two-cycle t_i. The plane of the physical page represents the four-cycle where the internal legs of D7, the two-cycle t_i and the point-like internal component of the D3-brane lie. Right: Representation of the mechanism that allows the interchange of winding modes between two D7-branes. These modes are wrapped on a two-dimensional submanifold characterised by t_i at the intersection of the D7-branes

$$\delta K_{g_s} = \sum_i g_s \frac{C_i(U,\bar{U}) t_i^{\perp}}{\mathcal{V}} + \sum_i \frac{\tilde{C}_i(U,\bar{U})}{t_i^{\cap}\mathcal{V}}, \tag{7.31}$$

where we can see that in the case of a toroidal geometry the volume dependence reduces to $\mathcal{V}_{T^6} \sim t^3 \sim t\tau$ and the corrections take the form of Eq. (7.30).

We would like to remark that the coefficients C_i, $\tilde{C}_i$ are, in general, unknown. However, since these depend only on the complex structure sector, we can take them to be constant at the perturbative level since the complex structure sector is fixed by tree-level fluxes. Similarly, we have written some dependence on the string coupling g_s which, a priori, depends on the vev of the axio-dilaton S. Taking the axio-dilaton fixed at tree-level by fluxes, we treat g_s as a constant at this level.

The scaling with the volume is a good educated guess of [28]. It is unclear how terms that scale with individual four-cycles (like in the case of the toroidal orientifold) could appear for smooth Calabi-Yaus. Although this idea cannot be completely disregarded, in all well-known examples where this computations can be carried out the volume dependence is the only one to appear. For example, in [28] the exact one-loop corrections are calculated for the quintic $\mathbb{P}_4^{[1,1,1,6,9]}$ and are found to agree with Eq. (7.3).

Recently, an important advancement was made to better understand the origin of loop corrections in generic Calabi-Yaus [30]. These results differ from those of [27] in the relative scaling of the winding and KK modes with the string coupling g_s. More importantly, the authors also flag the possible existence of log-enhanced corrections that could dominate the loop contributions. These log-corrections could arise from $\mathcal{R}^4$ operators localised on D7/O7 stacks or $\mathcal{R}^3$ localised on a 6D submanifold. A string amplitude calculation to confirm the existence of such terms is yet to be done.

Models that rely on loop corrections to lift flat directions, such as large volume scenario models with multiple Kähler moduli, could be in trouble if the existence of these operators were to be confirmed.

7.5 Higher Derivative Corrections

The last set of perturbative corrections that we would like to discuss are the higher derivative corrections of [31]. These corrections are intimately related to the ones in Sect. 7.3. Indeed, in that section we found that the Kähler potential gains a perturbative corrections such that

$$K \to K_0 - 2\ln\left(\mathcal{V} + \frac{\hat{\xi}}{2}\right) . \tag{7.32}$$

These corrections will then contribute to the four-derivative expansion of the scalar potential as follows. In [32], it was shown that the Ricci tensor along the internal components in the presence of α'-corrections is given by

$$R_{i\bar{j}} \sim (\alpha')^3 \partial_i \partial_{\bar{j}} \mathbb{S} , \tag{7.33}$$

where $\mathbb{S}$ was defined in Eq. (7.27). A priori, the metric is not Ricci-flat anymore and can be expanded as

$$g_{i\bar{j}} = g_{i\bar{j}}^{(0)} + (\alpha')^3 g_{i\bar{j}}^{(1)} , \tag{7.34}$$

where $g_{i\bar{j}}^{(0)}$ is the original Ricci-flat metric in the absence of corrections and $g_{i\bar{j}}^{(1)}$ solves the Einstein equations to $\mathcal{O}(\alpha'^3)$. At leading order in α', the only contribution from $g_{i\bar{j}}^{(1)}$ enters through the dimensional reduction of the ten-dimensional Ricci scalar. However, this turns out to be a total derivative meaning that the first contribution from $g_{i\bar{j}}^{(1)}$ comes at $\mathcal{O}(\alpha'^6)$. Therefore, at $\mathcal{O}(\alpha'^3)$ we can self-consistently ignore $g_{i\bar{j}}^{(1)}$ and consider $g_{i\bar{j}} = g_{i\bar{j}}^{(0)}$ to be Ricci-flat.

This implies that, at the four-dimensional effective level and at $\mathcal{O}(\alpha'^3)$, there are no curvature contributions coming from the internal piece. In the large volume limit and up to the four-derivative expansion, the corrections were calculated in [31] to be given at leading order in the $\mathcal{O}(1/\mathcal{V})$ expansion by Eq. (7.4)

$$\delta V = -g_s^{-3/2} \frac{3^4 \lambda W_0^4}{\mathcal{V}^4} \Pi_i t^i , \tag{7.35}$$

where Π_i are defined as in Eq. (7.5)

7.6 Non-perturbative Corrections

Finally, we will consider non-perturbative corrections due to instantons in the theory. Below, we introduce the origin of instanton corrections in relation with type IIB model building. A complete overview of instanton effects on type II theories can be found at [33].

To obtain an $\mathcal{N} = 1$ supersymmetric low energy effective theory from type IIB supergravity, we have seen that we need to introduce O_p-planes to project out half of supersymmetry generators. In particular, we introduce a holomorphic involution (σ) that acts on the Kähler form and holomorphic 3-form as

$$\sigma^* J \to J\,, \qquad \sigma^* \Omega \to \Omega\,. \tag{7.36}$$

The dimensionality of the orientifold is fixed by the dimensionality of the fix point set induced by σ. Taking $\Omega \propto dz^1 \wedge dz^2 \wedge dz^3$, for some set of $\{z^i, \bar{z}^i\}$ complex coordinates spanning the Calabi-Yau, we see that the condition $\sigma^*\Omega \to \Omega$ leads to fixed points[5] with complex dimensions 0 or 2, i.e. either points or complex planes. Given that the orientifold planes have to fill the external space to preserve four-dimensional Poincaré invariance, this particular choice of involution leads to $O3/O7$-planes.

To preserve the orientifolding, the instantons will have to be related to objects wrapping points or four-cycles. This gives rise to two possible contributions: one from Euclidean D3-branes and another one from gaugino condensation in D7-branes [34]. In [33], it was shown that, for the Euclidean D3-brane instantons, the non-perturbative corrections to the superpotential takes the schematic form

$$\begin{aligned} W(S,U) = W_0(S,U) \quad &\to \quad W(S,U,T) \\ &= W_0(S,U,\mathrm{e}^{-T}) + \sum_{E3_i} A_i(U)\mathrm{e}^{-a_i T} + A_S(U)\mathrm{e}^{-S}\,, \end{aligned} \tag{7.37}$$

where $E3_i$ denotes a stack of Euclidean D3-branes and (S, U, T) are the axio-dilaton, complex structure and Kähler chiral superfields, respectively. Noting that (S, U) are fixed at tree-level by fluxes, we find an expansion in non-perturbative corrections of the form in Eq. (7.6)

$$W(S,U,T) \simeq W_0(\langle S\rangle, \langle U\rangle) + \sum_{E3_i} A_i(\langle U\rangle)\mathrm{e}^{-a_i T}\,, \tag{7.38}$$

where $(\langle S\rangle, \langle U\rangle)$ are the vevs of the axio-dilaton and complex moduli sector, and we take W_0 and A_i constant at the level of the low-energy supergravity description. The discussion for the D7-brane contribution is quantitatively similar and the coefficient

[5] Submanifolds of a CY with one complex dimensions have to be given by a combination of holomorphic and antiholomorphic coordinates, as $H^2(\mathcal{X}) = H^{1,1}(\mathcal{X})$, and thus these are not invariant under σ.

a_i in the exponential is related to the rank N_i of the condensing gauge group on each stack of D7-branes, i.e. $a_i = 2\pi/N_i$.

We note that a third instanton contribution to the non-perturbative superpotential could arise from Euclidean D(-1)-branes. These effects would be the string theoretical analogue to classical localised solutions to the Euclidean equations of motion in QFT. We expect their contribution to the superpotential to be exponentially suppressed at weak coupling $W_{\text{ED}(-1)} \sim \mathcal{O}(\text{e}^{-1/g_s})$ [35] and thus will not consider them for model building purposes.

7.7 Validity of the Effective Description

Since we will be treating the supergravity models at an effective four-dimensional level to study their phenomenology, it is good to keep in mind the consistency requirements to truncate the effective theory to the set of corrections given above.

Tree-level hierarchy with respect to corrections. Most of the work in the subsequent sections and chapters will assume that the axio-dilaton and complex structure moduli have been stabilised at tree-level by fluxes. This allows us to concentrate on lifting the Kähler sector without having to worry about destabilising the other directions. To ensure that the other sectors are not destabilised we have to demand that the corresponding saxions are fixed at large vacuum expectation values by the fluxes.

For the dilaton, this is of two-fold importance as it also guarantees that the loop expansion is well-posed, as we will remark below. For the complex structure moduli, this allows us to truncate the prepotential to the *large complex structure term*, the equivalent to the leading term of Eq. (7.13) for the complex sector.

For the α' expansion. We already made use of a consistency check in Chap. 6, namely that the α' expansion is under control when the string frame volume $\mathcal{V}_S \gg 1$, this implied $\mathcal{V}/s \gg 1$ for the volume in Einstein frame. A second, but related, check is that consistency in truncating the α' expansion to the leading term requires

$$\frac{\hat{\xi}}{g_s^{3/2}\mathcal{V}} = \frac{\hat{\xi}}{\mathcal{V}_S} \ll 1 \,. \tag{7.39}$$

Indeed, further subleading terms carry extra derivatives with respect to the overall volume modulus. We expect these to be further suppressed in a $1/\mathcal{V}$ expansion as compared to the leading order α'^3-corrections. Thus, requiring that the α'^3 term satisfies Eq. (7.39) guarantees the safety of the truncation to leading order α'-corrections.

For the loop expansion. Given that the loop corrections are given to leading order in an expansion on the string coupling g_s, we require that the vev of the dilaton is fixed at large values

$$\frac{1}{g_s} := \text{e}^{-\langle\phi\rangle} = \langle \text{Re}(S) \rangle \gg 1 \,. \tag{7.40}$$

We can see that this consistency check comes *for free* by requiring that there exists a hierarchy between the tree-level actions and its corrections.

For the expansion in non-perturbative effects. For the instanton expansion to be well-posed we require that

$$e^{-a_i^a \operatorname{Re}(T_a)} \ll 1 , \qquad a_i^a \operatorname{Re}(T_a) \gg 1 \tag{7.41}$$

for all (i, a), where i labels the stack of Euclideanised D3-branes and $a = 1, ..., h^{1,1}$. In the large volume scenario described in Sect. 7.10, we will often work with only the leading term in this expansion, i.e. the smallest four-cycle $\operatorname{Re}(T_s)$, which implies a further hierarchy

$$1 \ll a_i^s \operatorname{Re}(T_s) \ll a_i^b \operatorname{Re}(T_b) , \tag{7.42}$$

where the label "b" indicates any four-cycle other than $\operatorname{Re}(T_s)$.

Supergravity scales. Together with the moduli fields a number of Kaluza-Klein degrees of freedom will formally appear in the effective four-dimensional description. We are interested in integrating these out so that the complete hierarchy of scales appearing in the effective description has to look like

$$M_p > M_s > M_{KK}^{(i)} \gg m_i, m_{3/2} , \tag{7.43}$$

where M_s is the string scale mass, $M_{KK}^{(i)}$ are the Kaluza-Klein masses that can appear from bulk modes or associated to D7-branes wrapping 4-cycles, m_i denotes the masses of the moduli fields and $m_{3/2}$ is the gravitino mass. It can be shown that these hierarchy can be made consistent when the other previous perturbative and non-perturbative checks are satisfied [36]. Indeed, this is so due to the volume scaling for the different masses

$$M_s \sim \frac{1}{\sqrt{\mathcal{V}}} , \qquad M_{KK} \sim \frac{1}{\mathcal{V}^{2/3}} , \qquad m_{3/2} \sim \frac{1}{\mathcal{V}} . \tag{7.44}$$

There will also be a number of phenomenological conditions on these scales (and others) due to observations (or lack thereof), but we will discuss them in the next chapter when they become more relevant.

With all this in mind we move on to discuss particular realisations of de Sitter vacua within supergravity. In general, the scale of the scalar potential will be set by some combination of the form $|W_0|^p / \mathcal{V}^q$, where $p, q > 0$. Obtaining a small enough scale to fit cosmological observations allows a natural categorisation of the model building efforts into two classes: the ones that require $|W_0| \ll 1$, like KKLT [10] and racetrack type scenarios [11, 37], and the ones requiring $\mathcal{V} \gg 1$, like the aptly named large volume scenario [29].

7.8 KKLT Model Building

To exemplify the KKLT scenario, we will consider a single Kähler modulus geometry where the volume of the Calabi-Yau is of the form $\mathcal{V} = (T + \bar{T})^{3/2}$, with $T := \tau + i\theta$ the single 4-cycle volume modulus. The key ingredient to generate a minimum for the Kähler direction in the KKLT scenario is the use of the non-perturbative corrections in Eq. (7.6) and the leading α' corrections in Eq. (7.2). In this case, the superpotential and Kähler potential are given by

$$W = W_0 + A\,\mathrm{e}^{-aT}\,, \qquad K = K_0 - 2\ln\left(\mathcal{V} + \frac{\hat{\xi}}{2}\right). \tag{7.45}$$

We note that neglecting loop and higher derivative contributions to the potential can be done consistently due to the volume hierarchy in Eq. (7.11) and that we will still require $\mathcal{V} \gg 1$. The dynamics of the moduli, to leading order in $\hat{\xi}$, is given by the Lagrangian

$$\mathcal{L} = K_{T\bar{T}}\,\partial T \partial \bar{T} - V(\tau,\theta) = \frac{3}{4\tau^2}\left(1 - \frac{5\hat{\xi}}{2^{7/2}\tau^{3/2}}\right)\left[(\partial\tau)^2 + (\partial\theta)^2\right] - V(\tau,\theta)\,, \tag{7.46}$$

where the F-term potential is

$$V(\tau,\theta) = \frac{a^2A^2\mathrm{e}^{-2a\tau}}{6\tau}\left(1 + \frac{3}{a\tau}\right) - \frac{aA|W_0|\mathrm{e}^{-a\tau}}{2\tau^2}\cos(a\theta) + \frac{3W_0^2\hat{\xi}}{64\sqrt{2}\tau^{9/2}}\,, \tag{7.47}$$

and, without loss of generality, we have assumed W_0 to be real and negative, $W_0 = -|W_0|$. It then follows that a SUSY minimum exists at zeroth order in $\hat{\xi}$, located at $\theta = 0$ and $\tau = \tau_{\rm min}$, where the latter satisfies the following relation

$$|W_0| = A\,\mathrm{e}^{-a\tau_{\rm min}}\left(1 + \frac{2}{3}a\tau_{\rm min}\right) \simeq \frac{2}{3}Aa\tau_{\rm min}\,\mathrm{e}^{-a\tau_{\rm min}}\,. \tag{7.48}$$

We work in a regime where the α'^3 corrections can be consistently neglected in the vicinity of the minimum thanks to Eq. (7.39). At large volumes these corrections will induce a maximum in the potential as stressed in [38], which will be relevant to the discussion in Chap. 8. In the past these very same corrections have been used for uplifting the SUSY AdS minimum to Minkowski [39], a more extreme regime which we will not consider here. For the moment, however, we concentrate on the dynamics close to the minimum, neglecting the subleading effects of $\hat{\xi}$. The potential at the minimum is AdS, and given by

$$V_{\rm AdS} \equiv V(\tau_{\rm min}, 0) = -\frac{a^2A^2\mathrm{e}^{-2a\tau_{\rm min}}}{6\tau_{\rm min}} = -3\left(\frac{|W_0|}{\mathcal{V}_{\rm min}}\right)^2. \tag{7.49}$$

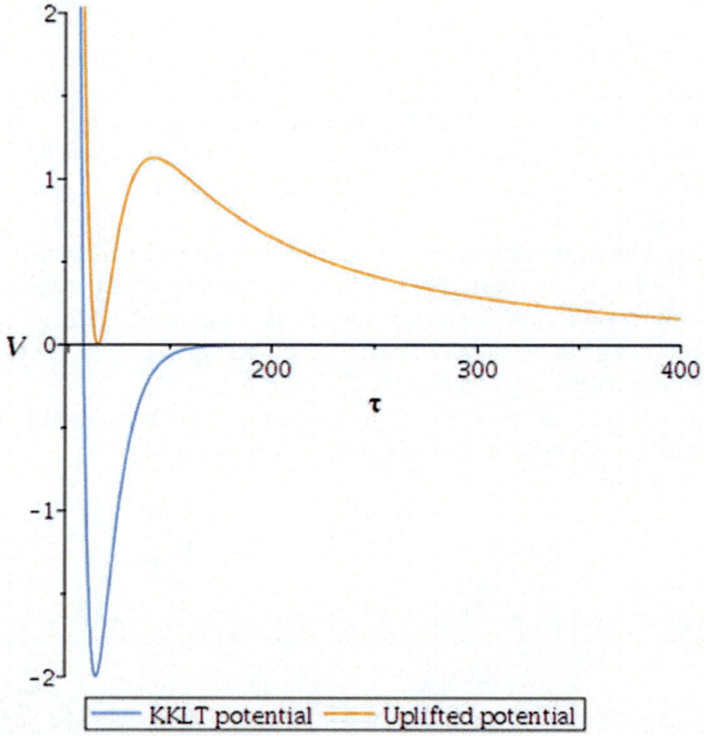

Fig. 7.2 KKLT scalar potential with and without uplift

When the axion is at its minimum, the corresponding KKLT potential for the saxion is given by $V_{\text{KKLT}}(\tau) = V(\tau, 0) + V_{\text{up}}(\tau)$. The uplift is tuned so that the new dS minimum, located at τ_{dS}, is compatible with current bounds on the cosmological constant, that is $V_{\text{KKLT}}(\tau_{\text{dS}}) = V(\tau_{\text{dS}}, 0) + V_{\text{up}}(\tau_{\text{dS}}) \lesssim 10^{-120}\, M_{\text{pl}}^4$. For the instanton expansion to be under control at the minimum, it must be placed at some large value of τ. As a consequence, the uplift does not have a huge effect on its position, and we can take $\tau_{\text{dS}} \simeq \tau_{\text{min}}$. It follows that the scale of the uplift is simply given by the scale of the AdS vacuum, as one might already have expected, $V_{\text{up}}(\tau_{\text{dS}}) \approx |V_{\text{AdS}}|$. Furthermore, since the original scalar potential decays exponentially quickly in comparison to the uplift term at large τ, the metastable vacuum is separated from the runaway region by a barrier whose height is fixed at the same scale, $V_* \sim V_{\text{up}}(\tau_{\text{dS}}) \approx |V_{\text{AdS}}|$. The generic shape of the potential with and without uplift is shown in Fig. 7.2.

It is clear now that, if the scale of the de Sitter vacuum is fixed by $V_{\text{AdS}} = 3\left(\frac{|W_0|}{\mathcal{V}_{\text{min}}}\right)^2$, to match the current observational bounds on the cosmological constant we require $W_0 \ll 1$, $\mathcal{V} \gg 1$ or a combination of small tree-level fluxed superpotential and large volume at the minimum. We also note that in this particular case with a single Kähler modulus, the volume and the fluxed superpotential are not independent. Indeed, from Eq. (7.48)

$$|W_0| \simeq \tau_{\rm min}\,{\rm e}^{-\tau_{\rm min}} \sim \mathcal{V}_{\rm min}^{2/3}{\rm e}^{-\mathcal{V}_{\rm min}^{2/3}}\,, \qquad V_{\rm dS} \simeq |V_{\rm AdS}| \sim \frac{{\rm e}^{-\mathcal{V}_{\rm min}^{2/3}}}{\mathcal{V}_{\rm min}}\,, \tag{7.50}$$

and with $V_{\rm dS} \sim 10^{-120}$ we find $\mathcal{V}_{\rm min} \sim \mathcal{O}(10^3)$ and $W_0 \sim \mathcal{O}(10^{-114})$. As noted before, this value of the volume at its minimum allows us to pass the consistency checks of Sect. 7.7 and provide a hierarchy in Eq. 8.7 so that subleading corrections can be safely ignored.

The usual lore goes that given that the string landscape could have as many as $\mathcal{O}(10^{272000})$ flux vacua[6] [43], albeit with a large degeneracy in the fluxes, finding such a small value for the fluxed superpotential should not be difficult. Recently, a mechanism to obtain a naturally small W_0 has been proposed in [44], where flux configurations where found so that $W_0 \lesssim \mathcal{O}(10^{-123})$.

Finally, we note that with supersymmetry now broken by the anti D3-brane, the gravitino acquires a mass at the uplifted minimum,[7] given by

$$m_{3/2}^2 = \left[{\rm e}^K|W|^2\right]_{\tau=\tau_{\rm dS},\theta=0} \approx \left[{\rm e}^K|W|^2\right]_{\tau=\tau_{\rm min},\theta=0} = \left(\frac{|W_0|}{\mathcal{V}_{\rm min}}\right)^2 = \frac{|V_{\rm AdS}|}{3} \tag{7.51}$$

where we have used the fact that $D_i W = 0$ and so $V_{\rm AdS} = -3{\rm e}^K|W|^2$ for the supersymmetric AdS vacuum.

As was noted in [45], the KKLT scenario suffers from the so-called Kallosh-Linde (KL) problem and a solution in the form of a racetrack type superpotential was offered. We will discuss this problem in detail in the following chapter and we continue by introducing the racetrack construction. At this point, suffice it to say that the issue is related to fact that the gravitino mass and the potential barrier are of the same order, since both are generated at the same order in the corrections.

7.9 Racetrack Model Building

The racetrack superpotential [11, 37] receives a second instanton contribution

$$W = W_0 + A\,{\rm e}^{-aT} + B\,{\rm e}^{-bT}\,, \tag{7.52}$$

as one would expect from gaugino condensation in a theory with a product gauge group.[8] In particular, for $SU(M) \times SU(N)$ we expect $a = 2\pi/M$ and $b = 2\pi/N$.

---

[6] This was originally computed using the techniques from the works of Ashok, Denef and Douglas [40–42].

[7] We remark that the precise uplifting mechanism is unimportant at this stage as supersymmetry must be broken by any uplift and the quantitative discussion about the gravitino mass follows through.

[8] This is equivalent to considering instanton corrections generated by two stacks of N and M Euclidean D3-branes.

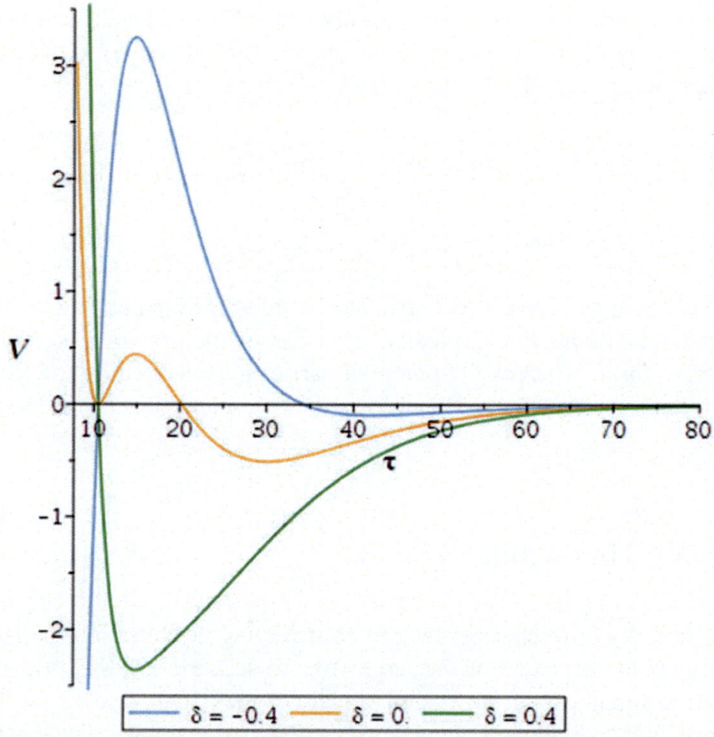

Fig. 7.3 Racetrack potential at $\theta = 0$ for parameter choice $A = 1$, $B = -1$, $a = 0.1$, $b = 0.09$ and different values of $W_0 = W_0^{\rm crit}(1+\delta)$

The corresponding F-term potential is given by

$$V(\tau,\theta) = \frac{a^2A^2\mathrm{e}^{-2a\tau}}{6\tau}\left(1+\frac{3}{a\tau}\right) + \frac{b^2B^2\mathrm{e}^{-2b\tau}}{6\tau}\left(1+\frac{3}{b\tau}\right)$$
$$+\frac{abAB\mathrm{e}^{-(a+b)\tau}}{3\tau}\left(1+\frac{3}{2a\tau}+\frac{3}{2b\tau}\right)\cos[(a-b)\theta]$$
$$-\frac{aA|W_0|\mathrm{e}^{-a\tau}}{2\tau^2}\cos(a\theta) - \frac{bB|W_0|\mathrm{e}^{-b\tau}}{2\tau^2}\cos(b\theta)\,. \tag{7.53}$$

In contrast with KKLT, the racetrack model admits a SUSY Minkowski vacuum for a critical value of W_0, given by

$$|W_0|^{\rm crit} = A\,\mathcal{R}^{\frac{a}{b-a}} + B\,\mathcal{R}^{\frac{b}{b-a}} \tag{7.54}$$

where we assume $\mathcal{R} = -\frac{aA}{bB} > 1$ and $a > b$ for definiteness. The minimum is located at $\theta = 0$ and $\tau_{\rm min} = \frac{1}{(a-b)}\ln\mathcal{R}$. The gravitino mass vanishes at the Minkowski vacuum since supersymmetry remains unbroken. The shape of the potential for different values of W_0 is shown in Fig. 7.3.

In this case, the tuning to exponentially small W_0 is alleviated in exchange for some extra tuning between the values of the coefficients (a, b) in the exponentials of the superpotential. Indeed, taking as an example the numbers in [45]

$$A = 1\,, \qquad B = -1.03\,, \qquad a = \frac{2\pi}{100}\,, \qquad b = \frac{2\pi}{99}\,, \tag{7.55}$$

yields $W_0^{\rm crit} \sim -10^{-4}$ and $\mathcal{V} \sim \mathcal{O}(1000)$. Obtaining a de Sitter vacuum now follows analogously to KKLT. Introducing an uplift term $\sim \frac{C}{\tau^2}$ will lead to a de Sitter vacuum and it is assumed that the C coefficient can be fine-tuned to obtain a viable scale for the cosmological constant. Critically, the mass of the gravitino is now generated by the small uplift whereas the potential barrier is generated by the leading order non-perturbative corrections. This allows the racetrack model to side-step the KL problem.

7.10 LVS Model Building

A fundamentally different approach to constructing de Sitter in type IIB effective theories goes by the name of the large volume scenario (LVS) [29]. In this case, the required smallness of the cosmological constant scale is obtained through an exponentially large internal volume with an $\mathcal{O}(1)$ value for W_0. The key ingredients for the LVS are an internal geometry containing large and small four-cycles and the introduction of loop and higher derivatives corrections to the theory.

In [29], it was shown that an AdS minimum can be generated in the large volume limit of type IIB supergravity in the presence of α' corrections and non-perturbative effects. We review the discussion below [46]. Consider a Calabi-Yau $\mathcal{X}$ whose Kähler sector is given by a number of moduli fields

$$\begin{cases} \tau_i\,, & \text{which remain small } \forall i = 1, ..., N_s\,, \\ \mathcal{V} \to \infty & \text{for } \tau_j \to \infty\,, \qquad \forall j = N_s + 1, ..., h^{1,1}(\mathcal{X})\,, \end{cases} \tag{7.56}$$

for a type IIB $\mathcal{N} = 1$ four dimensional effective theory whose Kähler potential and superpotential are given by

$$\begin{cases} K = K_0 - 2\ln\left(\mathcal{V} + \frac{\hat{\xi}}{2}\right)\,, \\ W = W_0 + \sum_{i \le N_s} A_i \mathrm{e}^{a_i T_i}\,. \end{cases} \tag{7.57}$$

Then the four-dimensional scalar potential admits non-SUSY AdS vacua at exponentially large volume $\mathcal{V} \sim \mathrm{e}^{a\tau_i}$ $(\forall i \le N_s)$ along $h^{1,1}(\mathcal{X}) - N_s - 1$ directions corresponding to the τ_j blow-up modes if and only if $\hat{\xi} > 0$ and τ_j resolves a point-like singularity, i.e. $K_{jj}^{-1} \sim \mathcal{V}\sqrt{\tau_j}$.

Let us consider a minimal example in which the volume of the Calabi-Yau is given by $\mathcal{V} = \tau_b^{3/2} - \tau_s^{3/2}$ where τ_s is a small cycle and τ_b is a blow-up mode, as defined in Eq. (7.56). The four-dimensional scalar potential when considering this geometry and Eq. (7.57) is given at leading order by

$$V = \frac{8a^2 A^2 \sqrt{\tau_s} e^{-2a\tau_s}}{3\mathcal{V}} - \frac{4\tau_s}{\mathcal{V}^2} a A e^{-a\tau_s} |W_0| \cos(a\theta) + \frac{3W_0^2 \hat{\xi}}{4\mathcal{V}^3} \tag{7.58}$$

where we have set $K_0 = 1$ and taken $W_0 < 0$ without loss of generality. The axion is stabilised at $\theta_* = 0$ and we have dropped subleading $e^{-b\tau_b}$ contributions in anticipation that the big cycle will be stabilised at $1 \ll a\tau_{s,*} \ll b\tau_{b,*}$. Extremising with respect to the volume and the small cycle modulus we find the well-known LVS results

$$\mathcal{V} = \frac{3\sqrt{\tau_{s,*}}\,|W_0|}{4aA} e^{a\tau_{s,*}} f\left(\tau_{s,*}\right), \quad f\left(\tau_{s,*}\right) := \frac{1 - \frac{1}{a\tau_{s,*}}}{1 - \frac{1}{4a\tau_{s,*}}}, \tag{7.59}$$

$$\frac{\hat{\xi}}{2} = \tau_{s,*}^{3/2} \frac{f\left(\tau_{s,*}\right)\left[4 - f\left(\tau_{s,*}\right)\right]}{3} = \left[\frac{1}{b} \ln\left(\frac{\mathcal{V}}{|W_0|}\right)\right]^{3/2}. \tag{7.60}$$

The minimum of the potential is a non-SUSY AdS vacuum

$$\langle V \rangle = -\frac{3\hat{\xi} W_0^2}{2\mathcal{V}^3}\left[\frac{1 - f(\tau_{s,*})}{4 - f(\tau_{s,*})}\right] \simeq -\frac{3\hat{\xi} W_0^2}{8a\tau_{s,*}\mathcal{V}^3}. \tag{7.61}$$

Once again, the introduction of an uplift mechanism to obtain de Sitter or Minkowski vacua is necessary.

Here we have seen the LVS in action. Even though the volume becomes exponentially large, the presence of the small cycle τ_s means that the non-perturbative contributions can compete against the leading α'-corrections generating a minimum along the τ_s direction. We remark that, in the presence of $h^{1,1}$ Kähler moduli the leading order scalar potential would only fix N_s directions and overall the volume modulus, leaving $h^{1,1} - N_s - 1$ directions flat that could be lifted by subleading perturbative effects, like the loop corrections or higher derivative corrections introduced in the earlier sections.

Now that we have a handle on how model building in the bulk of moduli space works, in the next chapter we would like to answer the question of what are the necessary ingredients of a successful quintessence model.

References

1. Cicoli M, De Alwis S, Maharana A, Muia F, Quevedo F (2019) De Sitter vs Quintessence in string theory. Fortsch Phys 67:1800079. https://doi.org/10.1002/prop.201800079, arxiv:https://arxiv.org/abs/1808.08967
2. Cicoli M, De Alwis S, Maharana A, Muia F, Quevedo F (2019) De Sitter vs Quintessence in string theory. Fortsch Phys 67:1800079. https://doi.org/10.1002/prop.201800079, arxiv:https://arxiv.org/abs/1808.08967
3. AbdusSalam S, Abel S, Cicoli M, Quevedo F, Shukla P (2020) A systematic approach to Kähler moduli stabilisation. JHEP 08:047. https://doi.org/10.1007/JHEP08(2020)047, arxiv:https://arxiv.org/abs/2005.11329
4. Cicoli M, Quevedo F, Savelli R, Schachner A, Valandro R (2021) Systematics of the α' expansion in F-theory. JHEP 08:099. https://doi.org/10.1007/JHEP08(2021)099, arxiv:https://arxiv.org/abs/2106.04592
5. Becker K, Becker M, Haack M, Louis J (2002) Supersymmetry breaking and alpha-prime corrections to flux induced potentials. JHEP 06:060. https://doi.org/10.1088/1126-6708/2002/06/060, arxivhttps://arxiv.org/abs/hep-th/0204254
6. Berg M, Haack M, Pajer E (2007) Jumping through loops: on soft terms from large volume compactifications. JHEP 09:031. https://doi.org/10.1088/1126-6708/2007/09/031, arxiv:https://arxiv.org/abs/0704.0737
7. Cicoli M, Conlon JP, Quevedo F (2008) Systematics of string loop corrections in type IIB Calabi-Yau flux compactifications. JHEP 01:052. https://doi.org/10.1088/1126-6708/2008/01/052, arxiv:https://arxiv.org/abs/0708.1873
8. Ciupke D, Louis J, Westphal A (2015) Higher-derivative supergravity and moduli stabilization. JHEP 10:094. https://doi.org/10.1007/JHEP10(2015)094, arxiv:https://arxiv.org/abs/1505.03092
9. Blumenhagen R, Cvetic M, Kachru S, Weigand T (2009) D-brane instantons in type II orientifolds. Ann Rev Nucl Part Sci 59:269. https://doi.org/10.1146/annurev.nucl.010909.083113, arxiv:https://arxiv.org/abs/0902.3251
10. Kachru S, Kallosh R, Linde AD, Trivedi SP (2003) De Sitter vacua in string theory. Phys Rev D 68:046005. https://doi.org/10.1103/PhysRevD.68.046005, arxiv:[https://arxiv.org/abs/hep-th/0301240
11. Blanco-Pillado JJ, Burgess CP, Cline JM, Escoda C, Gomez-Reino M, Kallosh R et al (2004) Racetrack inflation. JHEP 11:063. https://doi.org/10.1088/1126-6708/2004/11/063, arxiv:https://arxiv.org/abs/hep-th/0406230
12. Kallosh R, Linde AD (2004) Landscape, the scale of SUSY breaking, and inflation. JHEP 12:004. https://doi.org/10.1088/1126-6708/2004/12/004, arxiv:https://arxiv.org/abs/hep-th/0411011
13. Balasubramanian V, Berglund P, Conlon JP, Quevedo F (2005) Systematics of moduli stabilisation in Calabi-Yau flux compactifications. JHEP 03:007. https://doi.org/10.1088/1126-6708/2005/03/007, arxiv:https://arxiv.org/abs/hep-th/0502058
14. Conlon JP, Quevedo F, Suruliz K (2005) Large-volume flux compactifications: Moduli spectrum and D3/D7 soft supersymmetry breaking. JHEP 08:007. https://doi.org/10.1088/1126-6708/2005/08/007, arxiv:https://arxiv.org/abs/hep-th/0505076
15. Cicoli M, Conlon JP, Quevedo F (2008) General analysis of LARGE volume scenarios with string loop moduli stabilisation. JHEP 10:105. https://doi.org/10.1088/1126-6708/2008/10/105, arxiv:[https://arxiv.org/abs/0805.1029 0805.1029]
16. Grisaru MT, van de Ven AEM, Zanon D (1986) Two-dimensional supersymmetric sigma models on Ricci flat kahler manifolds are not Finite Nucl Phys B 277:388 . https://doi.org/10.1016/0550-3213(86)90448-7
17. Witten E (1986) New issues in manifolds of SU(3) holonomy. Nucl Phys B 268:79. https://doi.org/10.1016/0550-3213(86)90202-6

18. Burgess CP, Escoda C, Quevedo F (2006) Nonrenormalization of flux superpotentials in string theory. JHEP 06:044. https://doi.org/10.1088/1126-6708/2006/06/044, arxiv:https://arxiv.org/abs/hep-th/0510213
19. Berg M, Haack M, Kors B (2005) Loop corrections to volume moduli and inflation in string theory. Phys Rev D 71:026005. https://doi.org/10.1103/PhysRevD.71.026005, arxiv:https://arxiv.org/abs/hep-th/0404087
20. Polchinski J (2007) String theory, vol 2: superstring theory and beyond. Cambridge Monographs on Mathematical Physics, Cambridge University Press. https://doi.org/10.1017/CBO9780511618123
21. Becker K, Becker M, Haack M, Louis J (2002) Supersymmetry breaking and alpha-prime corrections to flux induced potentials. JHEP 06:060. https://doi.org/10.1088/1126-6708/2002/06/060, arxiv:https://arxiv.org/abs/hep-th/0204254
22. Candelas P, De La Ossa XC, Green PS, Parkes L (1991) A pair of Calabi-Yau manifolds as an exactly soluble superconformal theory. Nucl Phys B 359:21. https://doi.org/10.1016/0550-3213(91)90292-6
23. Ferrara S, Sabharwal S (1990) Quaternionic manifolds for type II superstring Vacua of Calabi-Yau Spaces. Nucl Phys B 332:317. https://doi.org/10.1016/0550-3213(90)90097-W
24. Greene BR, Plesser MR (1990) Duality in Calabi-Yau moduli space. Nucl Phys B 338:15. https://doi.org/10.1016/0550-3213(90)90622-K
25. Candelas P, Green PS, Hubsch T (1990) Rolling among Calabi-Yau Vacua. Nucl Phys B 330:49. https://doi.org/10.1016/0550-3213(90)90302-T
26. Grimm TW (2005) The effective action of type II Calabi-Yau orientifolds. Fortsch Phys 53:1179. https://doi.org/10.1002/prop.200510253, arxiv:https://arxiv.org/abs/hep-th/0507153
27. Berg M, Haack M, Kors B (2005) String loop corrections to Kahler potentials in orientifolds. JHEP 11:030. https://doi.org/10.1088/1126-6708/2005/11/030, arxiv:https://arxiv.org/abs/hep-th/0508043
28. Berg M, Haack M, Pajer E (2007) Jumping through loops: on soft terms from large volume compactifications. JHEP 09:031. https://doi.org/10.1088/1126-6708/2007/09/031, arxiv:https://arxiv.org/abs/0704.0737
29. Balasubramanian V, Berglund P, Conlon JP, Quevedo F (2005) Systematics of moduli stabilisation in Calabi-Yau flux compactifications. JHEP 03:007. https://doi.org/10.1088/1126-6708/2005/03/007, arxiv:https://arxiv.org/abs/hep-th/0502058
30. Gao X, Hebecker A, Schreyer S, Venken G, Loops, local corrections and warping in the LVS and other Type IIB models. arxiv:https://arxiv.org/abs/2204.06009
31. Ciupke D, Louis J, Westphal A (2015) Higher-derivative supergravity and moduli stabilization. JHEP 10:094. https://doi.org/10.1007/JHEP10(2015)094, arxiv:https://arxiv.org/abs/1505.03092
32. Freeman MD, Pope CN (1986) Beta functions and superstring compactifications. Phys Lett B 174:48. https://doi.org/10.1016/0370-2693(86)91127-5
33. Blumenhagen R, Cvetic M, Kachru S, Weigand T (2009) D-brane instantons in type II orientifolds. Ann Rev Nucl Part Sci 59:269 https://doi.org/10.1146/annurev.nucl.010909.083113, arxiv:https://arxiv.org/abs/0902.3251
34. Baumann D, Dymarsky A, Kachru S, Klebanov IR, McAllister L (2010) D3-brane potentials from fluxes in AdS/CFT. JHEP 06:072. https://doi.org/10.1007/JHEP06(2010)072, arxiv:https://arxiv.org/abs/1001.5028
35. Kim M, D-Instanton superpotential in string theory. https://arxiv.org/abs/2201.04634
36. AbdusSalam S, Abel S, Cicoli M, Quevedo F, Shukla P (2020) A systematic approach to Kähler moduli stabilisation. JHEP 08:047. https://doi.org/10.1007/JHEP08(2020)047, arxiv:https://arxiv.org/abs/2005.11329
37. Blanco-Pillado JJ, Burgess CP, Cline JM, Escoda C, Gomez-Reino M, Kallosh R et al (2006) Inflating in a better racetrack. JHEP 09:002. https://doi.org/10.1088/1126-6708/2006/09/002, arxiv:https://arxiv.org/abs/hep-th/0603129
38. Conlon JP (2018) The de Sitter swampland conjecture and supersymmetric AdS vacua. Int J Mod Phys A 33:1850178. https://doi.org/10.1142/S0217751X18501786, arxiv:https://arxiv.org/abs/1808.05040

39. Westphal A (2007) de Sitter string vacua from Kahler uplifting. JHEP 03:102. https://doi.org/10.1088/1126-6708/2007/03/102, arxiv:https://arxiv.org/abs/hep-th/0611332
40. Ashok S, Douglas MR (2004) Counting flux vacua. JHEP 01:060. https://doi.org/10.1088/1126-6708/2004/01/060, arxiv:https://arxiv.org/abs/hep-th/0307049
41. Denef F, Douglas MR (2004) Distributions of flux vacua. JHEP 05:072. https://doi.org/10.1088/1126-6708/2004/05/072 arxiv:https://arxiv.org/abs/hep-th/0404116
42. Denef F, Douglas MR (2005) Distributions of nonsupersymmetric flux vacua. JHEP 03:061. https://doi.org/10.1088/1126-6708/2005/03/061, arxiv:https://arxiv.org/abs/hep-th/0411183
43. Taylor W, Wang Y-N (2015) The F-theory geometry with most flux vacua. JHEP 12:164. https://doi.org/10.1007/JHEP12(2015)164, arxiv:https://arxiv.org/abs/1511.03209
44. Demirtas M, Kim M, McAllister L, Moritz J, Rios-Tascon A (2021) Small cosmological constants in string theory. JHEP 12:136. https://doi.org/10.1007/JHEP12(2021)136, arxiv:https://arxiv.org/abs/2107.09064
45. Kallosh R, Linde AD (2004) Landscape, the scale of SUSY breaking, and inflation. JHEP 12:004. https://doi.org/10.1088/1126-6708/2004/12/004, arxiv:https://arxiv.org/abs/hep-th/0411011
46. Cicoli M (2010) String loop moduli stabilisation and cosmology in IIB flux compactifications. Fortsch Phys 58:115. https://doi.org/10.1002/prop.200900096, arxiv:https://arxiv.org/abs/0907.0665

Chapter 8
Obstructions to Quintessence Model Building

In this chapter, we consider the challenges in establishing the microscopic origin of both inflation and dark energy. In [1] it has been conjectured that scalar potentials that can be derived from putative quantum gravity theories obey the bound

$$V_\phi \geq \frac{c}{M_p} V, \tag{8.1}$$

where c is a positive and dimensionless order one constant. If true, this conjecture has serious implications for inflation in the early universe and dark energy at present times. The most obvious consequence is that de Sitter (dS) vacua are forbidden, ruling out the cosmological constant as the source of dark energy. However, the bound is also in some tension with the requirement of slow roll in two derivative scalar actions, both for inflation and dark energy. While this tension is stronger in the context of inflation, it may be acceptable for dark energy models given that current bounds on ω_{DE} [2] are more relaxed that those derived from the scalar spectral tilt, n_s, for inflationary models [3, 4]. It was later realised that this bound would rule out the experimentally tested Higgs potential, and would preclude electroweak symmetry breaking which requires $V_\phi = 0$ for $V_{\phi\phi} < 0$ and $V > 0$ [5]. Moreover, it would also rule out supersymmetric AdS vacua that are accompanied by dS maxima at large field values [6]. This unsatisfactory state of affairs prompted the proposal of a refined conjecture that took the form [7, 8]

$$V_\phi \geq \frac{c}{M_p} V \quad \text{or} \quad V_{\phi\phi} \leq -\frac{c'}{M_p^2} V, \tag{8.2}$$

where c and c' are positive and dimensionless order one constants. These conjectures are not based on rigorous proofs and several counterexamples have been proposed [9–18]. Rather, the logic behind their formulation is the theoretical difficulty in establishing the existence of a dS vacuum in a fully convincing manner, mainly due to the need to break supersymmetry. Strong evidence in favour of the refined version of the conjecture has been given in [8] for any parametrically controlled

F. Cunillera, *Dark Energy*, Springer Theses,
https://doi.org/10.1007/978-3-031-21468-4_8

regime of string theory using a combination of the distance conjecture and entropy considerations. This is the regime where the semiclassical approximation can be made arbitrarily good by sending the parameters that control the string loop and the α' expansions to zero. These are, respectively, the real part of the axio-dilaton S, which sets the string coupling $g_s = 1/\text{Re}(S)$, and the extra-dimensional volume in string units $\mathcal{V}$ which controls the α' expansion since $1/\mathcal{V}^{1/3} = \alpha'/\text{Vol}^{1/3}$ (where Vol is the dimensionful volume). The asymptotic limit where $\text{Re}(S) \to \infty$ and $\mathcal{V} \to \infty$ corresponds to the semiclassical approximation with no dS vacua.

However, dS vacua could still exist in the bulk of moduli space where the quality of the approximations should be carefully checked. In particular, a necessary condition to have control over the effective field theory is the existence of small expansion parameters such as the flux-generated superpotential $W_0 \ll 1$ in KKLT models [19] and the inverse of the internal volume $1/\mathcal{V} \ll 1$ in LVS vacua [20–22]. Much progress has been made in this direction by determining perturbative [23–29] and non-perturbative corrections [30], or by estimating their moduli dependence using higher dimensional arguments based on symmetries [31] and geometry [32]. However, it is fair to say that the existence of dS vacua in the interior of the moduli space has still to be established in a fully convincing manner and there are a growing number of no-go theorems explicitly demonstrating their absence in particular compactifications of string inspired effective theories [33–41]. Even if they *did* exist, dS vacua in string theory might well be short-lived, as suggested by the TCC conjecture [42]. Of course, this is not a problem for dark energy as observations only require it to be dominant for a single efolding of accelerated expansion. This may even be desirable in the context of the cosmological coincidence problem, as discussed in Chap. 5 [43–45].

All these considerations show that the existence of dS vacua in string theory is still an open problem which requires further scrutiny. It is, therefore, interesting to investigate if the alternative to a cosmological constant—namely, quintessence—shares the same technical difficulties. To this end, we shall focus on the microscopic origin of dark energy as a dynamically evolving scalar field emerging from a compactification of string theory.

First of all, let us mention that, similar to dS constructions, quintessence cannot be realised in any parametrically controlled regime of string theory since dilaton or volume mode runaways in the asymptotics of the moduli space are too steep to drive an epoch of accelerated expansion [46], see Chap. 6. The obstruction echoes some of the obstructions to dS vacua [8], with related results for quintessence also being derived in [47–49]. Note that the situation does not improve if one performs a multifield evolution including their corresponding axionic fields. In fact, even if non-geodesic trajectories on curved field manifolds could, in principle, yield a period of accelerated expansion for steep potentials [50, 51], this is never the case for either $\text{Re}(S)$ or $\mathcal{V}$ [52].

As a result, quintessence can only be realised in the bulk of moduli space where it generically shares the same control issues as dS model building [53]. On top of the technical difficulties in trusting the effective field theory, quintessence is known to feature some phenomenological challenges including the 'light volume problem' and the 'F-term problem' [54]. The 'light volume problem' relates to quintessence driven by a saxion, typically a volume modulus. To be compatible with the acceleration we see today, this modulus needs to be extremely light, with its mass bounded above by the current Hubble scale. As it also couples to matter with gravitational strength, this would yield an additional long range scalar force, in violation of fifth force constraints [55]. The 'F-term problem' is associated with radiative corrections involving supersymmetric particles running in loops, producing contributions to the scalar potential that are much larger than the dark energy scale. Traditional quintessence, at least in a perturbative regime, also has some observational problems, having been shown to enhance the so-called Hubble tension [56, 57] which is already at 5σ for ΛCDM [58].

Here we add to the challenges facing quintessence in string theory. In particular, we show how a version of the so called 'Kallosh-Linde (KL) problem' [59] drastically constrains the spectrum of possibilities. The KL problem is one of runaway behaviour in the volume mode during inflation. It is normally used to constrain the scale of inflation against the gravitino mass. We use it to constrain the form of the underlying scalar potential responsible for dark energy, exploiting the huge hierarchy of scales between the acceleration today and in the early universe. This hierarchy makes it extremely difficult to have a scalar potential that is compatible with current observations and is protected from the KL runaway during inflation.

Let us briefly run through the logic. We begin with $V_0(\mathcal{V})$, the potential that fixes the volume mode. However, the volume mode also couples to any source of energy-momentum thanks to the Weyl rescaling to four-dimensional Einstein frame. As a result, in Einstein frame, there is a direct coupling between $\mathcal{V}$ and the potentials for both the inflaton σ and the quintessence field ϕ. The total scalar potential describing the dynamics of all three fields is given by $V_{\text{tot}} = V_0(\mathcal{V}) + V_1(\sigma, \mathcal{V}) + V_2(\phi, \mathcal{V})$ where $V_1(\sigma, \mathcal{V})$ is generated from the inflaton potential and $V_2(\phi, \mathcal{V})$ from the quintessence potential. Recall that there exists an enormous hierarchy between the energy scales of inflation and dark energy: $V_{\text{inf}} \gtrsim (1\,\text{MeV})^4 \gg (1\,\text{meV})^4 \sim V_{\text{DE}}$.

During inflation, with σ in slow roll, it follows that the quintessence field ϕ should be frozen, with $V_1(\sigma, \mathcal{V}) \gg V_2(\phi, \mathcal{V})$. Furthermore, in order to avoid destabilising the volume direction [59], we need to impose the condition $|V_0(\mathcal{V}_*)| \gtrsim V_1(\sigma, \mathcal{V}) \gg V_2(\phi, \mathcal{V})$, where $\mathcal{V}_*$ is the value of $\mathcal{V}$ controlling the barrier against decompactification. For Minkowski vacua $\mathcal{V}_* = \mathcal{V}_{\text{max}}$, the value of $\mathcal{V}$ at the top of the barrier, while for AdS vacua $\mathcal{V}_* = \mathcal{V}_{\text{min}}$, the value of the $\mathcal{V}$ at the minimum.[1] Of course, for AdS vacua inflation is possible only if $V_1(\sigma, \mathcal{V})$ acts as an uplifting term such that $V_{\text{inf}} \simeq V_0(\mathcal{V}_{\text{min}}) + V_1(\sigma, \mathcal{V}_{\text{min}}) > 0$. (Here we are assuming that the location of the minimum, $\mathcal{V}_{\text{min}}$, does not change significantly in the presence of the uplift.)

After the end of inflation $V_1(\sigma, \mathcal{V})$ goes to zero and so $V_{\text{tot}} \simeq V_0(\mathcal{V}_{\text{min}}) + V_2(\phi, \mathcal{V}_{\text{min}})$. For the case where $V_0(\mathcal{V})$ admits a (near) Minkowski vacuum with $V_0(\mathcal{V}_{\text{min}}) \simeq 0$, it follows that $V_{\text{tot}} \simeq V_2(\phi, \mathcal{V}_{\text{min}}) \simeq V_{DE} \lesssim 10^{-36} V_{\text{inf}} \lesssim 10^{-36}$

[1] For dS vacua the story is slightly different: we need to impose $V_1(\sigma, \mathcal{V}) \lesssim (V_0(\mathcal{V}_{\text{max}}) - V_0(\mathcal{V}_{\text{min}}))$, although we shall ignore this case since quintessence model building is less well motivated in the presence of a dS vacuum.

$V_0(\mathcal{V}_{\max})$, implying a *huge* hierarchy between the energy scales associated with the potential that stabilises $\mathcal{V}$ and the one which drives quintessence. For the case where $V_0(\mathcal{V})$ admits an AdS vacuum, the hierarchy of scales ensures that, after inflation, $V_{\text{tot}} \simeq V_0(\mathcal{V}_{\min}) + V_2(\phi, \mathcal{V}_{\min}) \simeq V_0(\mathcal{V}_{\min}) < 0$, implying that quintessence model building is not possible.

Notice that similar considerations would apply if the volume also plays the rôle of the inflaton ($\sigma = \mathcal{V}$), notwithstanding that explicit constructions of volume inflation look rather contrived [60]. Alternatively, if the volume plays the rôle of quintessence ($\phi = \mathcal{V}$) its potential would, again, be destabilised by the inflationary energy density. Finally if the volume is everything ($\sigma = \phi = \mathcal{V}$), we would require the presence of two slow roll regions at hierarchically different field values. Given that plateau-like regions can be obtained only by balancing competing terms, if the quintessence epoch at large field values is under control, the inflationary era would lie in a region where perturbation theory would tend to break down. Reheating after the end of inflation and fifth force constraints would also present additional problems in this particular case.

These considerations can be combined with implications of the refined dS conjecture [8]. The refined dS conjecture rules out quintessence models with a very shallow potential, as in [61], but allows for quintessence rolling near a hilltop at positive energy (perhaps in the presence of a global AdS or supersymmetric Minkowski vacuum) or down an exponential potential of the form $V = V_0\, e^{-\lambda\phi/M_P}$. In the latter case, it has been shown that agreement with data requires $\lambda \leq 1.02$ at 3σ [62].[2]. However, our analysis suggests that these two scenarios are not under better control than dS vacua.

Exponential potentials arise from no-scale breaking perturbative effects for saxions and are typically not small enough to produce the required hierarchy in scales between inflation and dark energy. Therefore, these models are expected to be destabilised by the inflationary dynamics, as well as suffering from problems with the light volume and the F-term. The KL problem also applies to hilltop quintessence near a maximum at positive energy, with a global AdS minimum. We shall present explicit examples of these scenarios and elucidate their problems in Sect. 8.2.

To avoid the KL problem, we could consider hilltop quintessence models with a supersymmetric (near) Minkowski vacuum. However, in these models the gravitino mass would be of order the dark energy scale resulting in violation of current bounds [64–67]. Moreover, supersymmetric Minkowski solutions are highly constrained, requiring a very precise form for the superpotential W. Therefore, even if the presence of supersymmetry might seem a powerful tool to keep computational control over these solutions, proving their existence in the interior of the moduli space might still be a challenge. As an illustrative example, consider the well-understood type IIB compactifications with H_3 and F_3 flux, and a tree-level W that does not depend on the complex volume mode $T = \tau + i\theta$. In this set-up, any supersymmetric Minkowski solution at tree-level would necessarily feature a complex flat direction, given by T.

[2] A stronger bound of $\lambda \leq 0.6$ was obtained in [63]. We refer the reader to [62] for a discussion of the two approaches.

The existence of a global Minkowski solution with all moduli stabilised would, therefore, have to rely on the existence of non-perturbative corrections, which lack a full systematic understanding. They would also lift τ and θ at the same level of approximation, without generating the right hierarchy between the would-be quintessence field θ and the volume mode τ.

In the end, we arrive at a generic picture for building a viable quintessence model in string phenomenology. Let us summarize the main points:

- At leading order (in either perturbative or non-perturbative expansions), the scalar potential $V_0(\mathcal{V})$ should feature a (near) Minkowski vacuum with a stabilised volume mode. Notice that non-supersymmetric Minkowski vacua typically require the inclusion of uplifting sectors, and so look qualitatively similar to dS vacua. Although supersymmetric Minkowski solutions could give better computational control, the subdominant effects which generate dark energy would also be responsible for supersymmetry breaking. The gravitino mass (and the soft terms) would not be decoupled from the dark energy scale, in strong tension with both particle physics [64–66] and cosmological observations [67]. Thus the leading order Minkowski vacuum should be *non*-supersymmetric.
- At the leading order of approximation, the quintessence field should remain flat in order to be able to create the required hierarchy between $V_0(\mathcal{V}_{\text{max}})$ and V_{DE}, with the latter generated by subdominant contributions. The presence of a flat direction can be guaranteed by shift symmetries which fall into two categories: (1) non-compact rescaling symmetries for saxions arising from the underlying no-scale structure [68, 69]. However, these are broken by perturbative effects, and so are not generally efficient enough to provide the required hierarchy; and (2) compact shift symmetries for axions which can potentially generate huge hierarchies, being broken only by tiny non-perturbative effects. Moreover, the smallness of these non-perturbative corrections ensures that the energy density associated with the quintessence potential does not destabilise the volume minimum.
- Axion quintessence automatically avoids the fifth-force problem (being driven by a pseudo-scalar) and ensures radiative stability thanks to the fact that the axionic shift symmetry is exact at the perturbative level.
- The main problem with axion quintessence is that its potential is flat enough to drive a period of accelerated expansion only if the axion decay constant is trans-Planckian. However, this situation is very difficult to realise since explicit string constructions with control over the effective field theory tend to have axions with sub-Planckian decay constants [70], as also implied by the weak gravity conjecture [71]. There could be counter-examples based on alignment mechanisms [72, 73], although their trustability requires further scrutiny.
- For generic axion potentials with sub-Planckian decay constants, we might seek quintessence from a hilltop model. Even if this possibility looks attractive from a model building perspective, we shall see in Sect. 8.3 that, when combined with quantum diffusion during the inflationary epoch [74], it relies on two conditions: (1) *very* finely tuned initial conditions; and (2) an extremely low inflationary scale ($H_{\text{inf}} \lesssim 1\,\text{MeV}$), at least for axion decay constants in the regime where the effective field theory is under control.

In other words, from the point of view of theoretical and phenomenological control, quintessence model building in string theory is at least as challenging as the search for dS vacua.

This conclusion raises doubts over the validity of the swampland dS conjecture. Taken alongside the challenges to quintessence, it would imply strong tension between quantum gravity and observation. This might be an indication that phenomenologically relevant solutions to string theory, like dS vacua, lie in the bulk of the moduli space. In this case, it might still be true that perturbation theory is a valid approximation but to be confident of this, we need to refine our technical ability to compute quantum corrections. In the end Nature has already shown an affinity for couplings (as in standard gauge theories and cosmological perturbation theory) that are weak enough to allow us to describe it to a good approximation, even if they cannot be made arbitrarily small.

Finally, if data were to prefer dynamical dark energy, our analysis shows that quintessence models are very unlikely to be axion hilltops since they require highly tuned initial conditions and a very low Hubble scale during inflation. In this regard, axion quintessence models based on alignment mechanisms look more promising even if they need further studies to be convincingly established in fully fledged string compactifications with moduli stabilisation.

8.1 Old Challenges for Quintessence in String Theory

We begin by recalling some relevant results from Chap. 7. In particular, the scalar potential in the presence of the leading α', g_s, higher derivative and non-perturbative corrections is given by

$$V = \delta V_{\alpha'} + \delta V_{g_s} + \delta V_{\text{hd}} + \delta V_{\text{np}} \tag{8.3}$$

where δV_{hd} is given by Eq. (7.4) and (setting $e^{K_0} = 1$)

$$\delta V_{\alpha'} = \frac{3\hat{\xi} W_0^2}{4\mathcal{V}^3}, \tag{8.4}$$

$$\delta V_{g_s} = \frac{W_0^2}{\mathcal{V}^2} \sum_i \left(g_s^2 C_i^2 K_{ii}^{\text{tree}} - 2\frac{\tilde{C}_i}{\mathcal{V} t_i^{\cap}} \right), \tag{8.5}$$

$$\delta V_{\text{np}} = \sum_{i,j} K_{\text{tree}}^{ij} a_i a_j A_i A_j \frac{e^{-a_i T_i - a_j \bar{T}_j}}{\mathcal{V}^2} + \frac{2W_0}{\mathcal{V}^2} \sum_i A_i a_i \tau_i \left(e^{-a_i T_i} + e^{-a_i \bar{T}_i} \right) \tag{8.6}$$

In the regime where all 2-cycles scale as $t \sim \sqrt{\tau} \sim \mathcal{V}^{1/3}$, we have the scaling

$$\delta V_{\alpha'} \sim \frac{W_0^2}{\mathcal{V}^3}, \quad \delta V_{g_s} \sim \frac{W_0^2}{\mathcal{V}^{10/3}}, \quad \delta V_{\text{np}} \sim \frac{\mathcal{V}^{4/3} e^{-2a\tau} + W_0 \mathcal{V}^{2/3} e^{-a\tau}}{\mathcal{V}^2}, \quad \delta V_{\text{hd}} \sim \frac{W_0^4}{\mathcal{V}^{11/3}}, \tag{8.7}$$

where we have taken $K_{ij}^{\text{tree}} \sim 1/\mathcal{V}^{4/3}$. To generate stable vacua, one has to find a balance between different terms in the potential. For example, in KKLT models [19], W_0 is tuned to exponentially small values, $W_0 \sim (a\tau)e^{-a\tau} \ll 1$, so that the two contributions to δV_{np} are comparable in size. This typically yields a supersymmetric AdS vacuum whose depth is parametrised by $-W_0^2/\mathcal{V}^2$. Upon uplift the same scale controls the height of the barrier separating the vacuum from the decompactification limit [19]. A notable exception to this rule is the racetrack setup which we describe in Sect. 8.2.1 [59, 75], where the scale of the vacuum can be made arbitrarily small thanks to two instanton contributions that are aligned relative to one another.

Another possible approach is to balance perturbative against non-perturbative corrections. Generically we expect the latter to be suppressed, except in the presence of small cycles, as this raises the size of the instanton correction. This is precisely what happens in the LVS scenario [20–22], where $\delta V_{\alpha'} \sim \frac{W_0^2}{\mathcal{V}^3} \sim \delta V_{\text{np}}$. This sets the scale of the potential, controlling both the depth of the non-supersymmetric AdS vacuum and the height of the barrier to infinity which develops after uplifting.

8.1.1 Fifth Forces and Radiative Instability

Some dynamical dark energy models have already been built within the framework of string compactifications [61, 76–80]. Typically the quintessence field corresponds to the lightest mode and the other moduli are stabilised at tree-level and by leading order corrections. In this way dynamical dark energy appears as a next-to-leading order effect, allowing us to retain perturbative control. It also guarantees that the slow roll of the quintessence field away from the minimum only displaces the volume mode from its original vacuum expectation value by a small amount.

However none of the existing quintessence models in the literature is really satisfactory due to several challenges which were already highlighted in [53, 54]. These challenges are related to the phenomenological requirements that a prospective stringy quintessence field would have to satisfy, namely:

1. A *light quintessence modulus* ϕ with $m_\phi \lesssim H_0 \sim 10^{-60}\, M_p$. This follow directly from requiring that the scalar field ϕ is in slow roll at the current epoch.
2. *Heavy superpartners* with masses $M_{\text{soft}} \gtrsim 10^{-15}\, M_p$. Supersymmetric partners must be above the threshold set by the LHC [81]. This, in turn, yields large perturbative corrections from loops of visible sector supersymmetric particles.
3. *Heavy Kaluza-Klein scale* with $M_{\text{KK}} \gtrsim 10^{-30}\, M_p$. Sub-millimetre scale tests of Newtonian gravity put a bound on the Kaluza-Klein scale [82].
4. *Heavy volume modulus* with $m_{\mathcal{V}} \gtrsim 10^{-30}\, M_p$. Upon compactification, the four-dimensional Ricci scalar gets a prefactor which depends on the volume modulus which couples to matter fields after Weyl rescaling to Einstein frame. There are stringent bounds on such fifth force effects given by sub-millimetre experiments [82–84].

The authors of [54] discuss the implications of these requirements for string models of dark energy, with a focus on LVS-motivated scenarios for concreteness. Two main issues arise.

The light volume problem

The Kaluza-Klein mass is given by

$$M_{\rm KK} = \frac{M_s}{R} \sim \frac{M_p}{\mathcal{V}^{2/3}} \gtrsim 10^{-30}\, M_p \qquad \Rightarrow \qquad \mathcal{V} \lesssim 10^{45}, \tag{8.8}$$

where we have used $M_s \simeq M_p\, \mathcal{V}^{-1/2}$, the fact that the radius of the compact space $R \sim \mathcal{V}^{1/6}$, and the bound on the Kaluza-Klein mass given above.

In the LVS scenario, the mass for the volume modulus is generated through leading α' corrections Eq. (8.4), while at subleading order loop corrections Eq. (8.5) lift additional Kähler moduli which could play the rôle of the quintessence field ϕ. Using Eqs. (8.7) and (8.8), one finds

$$\frac{m_\phi}{m_{\mathcal{V}}} \sim \sqrt{\frac{\delta V_{g_s}}{\delta V_{\alpha'}}} \sim \frac{1}{\mathcal{V}^{1/6}} \gtrsim 10^{-7}\,. \tag{8.9}$$

In [61] loop contributions are suppressed due to low energy supersymmetry in the bulk and an anisotropic shape of the extra dimensions. The quintessence field ϕ is instead lifted by poly-instanton effects which give

$$\frac{m_\phi}{m_{\mathcal{V}}} \sim \sqrt{\frac{\delta V_{\rm poly}}{\delta V_{\alpha'}}} \sim \frac{1}{\sqrt{\mathcal{V}}} \gtrsim 10^{-22}\,. \tag{8.10}$$

However, both (8.9) and (8.10) are in contradiction with the phenomenological bound imposed by fifth force constraints and the value of H_0, i.e. $\frac{m_\phi}{m_{\mathcal{V}}} \lesssim 10^{-30}$. A way to avoid this issue is to introduce subleading effects that modify the volume scaling of Eq. (8.9) and (8.10) [85]. For example, in the model of [61], ϕ does not mediate any fifth force since its coupling to Standard Model fields is weaker than Planckian due to sequestering effects in the extra dimension (see also [84] for estimates of moduli couplings in sequestered models with large extra dimensions). Nevertheless, the volume mode would lead to new long range interactions since, due to (8.10), it is much lighter than 1 meV and it couples with ordinary matter with standard Planckian strength (however, see [86] for a possible screening effect due to the kinetic coupling of $\mathcal{V}$ to its associated axionic field).

The F-term problem

The mass of the superpartners, which we approximate by the gaugino mass, is of the order

$$M_{\rm soft} \sim M_{1/2} = \frac{F^i \partial_i f}{{\rm Re}(f)}\, M_p, \tag{8.11}$$

where f is the gauge kinetic function. If we assume that supersymmetry breaking is mediated through some higher-dimensional operator at some scale M_b, for a simple toy model with a single spurion field X and F-term F_X, the contribution to the scalar potential is

$$\delta V_X \sim F_X^2 \sim M_b^2 M_{\rm soft}^2, \tag{8.12}$$

where $M_{\rm soft}$ enters the scalar potential after canonical normalisation of the spurion term. If we require that supersymmetry breaking is mediated above the TeV scale, together with the phenomenological constraints on the superpartner masses, we find $\delta V_X \gtrsim 10^{-60} M_p \gg H_0^2$. This contribution would raise the scale of the potential well beyond the dark energy scale. A loophole is to consider a new contribution to the scalar potential that would cancel supersymmetry breaking effects with some fine-tuning, as in [61], where the additional effect is assumed to come from the backreaction of non-supersymmetric visible sector branes (see also [87] for recent developments of quintessence models in scenarios with non-linearly realised supersymmetry).

The challenges for quintessence outlined in [53, 54] are just the tip of the iceberg. In the next section, we identify an even bigger problem: disruption of the energetic dynamics by the inflationary energy density, resulting in destabilisation of the volume mode and decompactification.

8.2 The KL Problem for Quintessence

In [59] Kallosh and Linde argued that the scale of inflation is bounded from above by the gravitino mass in the standard KKLT scenario [19]. The constraint arises in order to avoid a runaway in the volume mode, leading to decompactification at early times. Similar considerations were used to place limits on thermal corrections to the scalar potential, imposing a maximum temperature in the four-dimensional effective theory [88, 89]. The KL problem extends beyond KKLT, and has also been shown to affect LVS models where the constraint turns out to be even stronger [90]. We begin by reviewing the key aspects of the original argument of [59]. Later we will show that it has implications also for string models of dynamical dark energy.

8.2.1 Review of the KL Problem

Consider the one-instanton KKLT model of Sect. 7.8 with superpotential and α'-corrected Kähler potential given by

$$W = W_0 + A\, e^{-aT}, \qquad K = K_0 - 2\ln\left(\mathcal{V} + \frac{\hat{\xi}}{2}\right), \tag{8.13}$$

where $\mathcal{V} = (T + \bar{T})^{3/2}$ is the volume of the internal Calabi-Yau manifold with $T = \tau + i\theta$ the 4-cycle volume modulus. The dynamics of the moduli, to leading order in $\hat{\xi}$, is given by the Lagrangian

$$\mathcal{L} = K_{T\bar{T}}\, \partial T \partial \bar{T} - V(\tau, \theta) = \frac{3}{4\tau^2}\left(1 - \frac{5\hat{\xi}}{2^{7/2}\tau^{3/2}}\right)\left[(\partial\tau)^2 + (\partial\theta)^2\right] - V(\tau, \theta), \tag{8.14}$$

where the F-term potential is

$$V_{\text{KKLT}}(\tau, \theta) = \frac{a^2 A^2 e^{-2a\tau}}{6\tau}\left(1 + \frac{3}{a\tau}\right) - \frac{aA|W_0|e^{-a\tau}}{2\tau^2}\cos(a\theta) + \frac{3W_0^2\hat{\xi}}{64\sqrt{2}\tau^{9/2}}. \tag{8.15}$$

and, without loss of generality, we have assumed W_0 to be real and negative, $W_0 = -|W_0|$. From this potential, we found SUSY AdS vacua located at

$$V_{\text{AdS}} \equiv V_{\text{KKLT}}(\tau_{\text{min}}, 0) = -\frac{a^2 A^2 e^{-2a\tau_{\text{min}}}}{6\tau_{\text{min}}} = -3\left(\frac{|W_0|}{\mathcal{V}_{\text{min}}}\right)^2. \tag{8.16}$$

which was uplifted via

$$V_{\text{up}} = \frac{C}{\tau^2}. \tag{8.17}$$

Furthermore, by demanding that after uplifting the minimum of this potential sat at $V_{\text{KKLT}'}(\tau_{\text{dS}}) = V_{\text{KKLT}}(\tau_{\text{dS}}, 0) + V_{\text{up}}(\tau_{\text{dS}}) \lesssim 10^{-120}\, M_p^4$, we obtained a relation between the mass of the gravitino $m_{3/2}$ and the height of the potential barrier, V_*, protecting the volume modulus from a runaway

$$m_{3/2}^2 = \frac{|V_{\text{AdS}}|}{3} \sim V_*\,. \tag{8.18}$$

The KL problem emerges when we consider inflation in this particular setup. To begin with, one could consider hilltop inflation from the top of the potential barrier V_*. In this case, the Hubble parameter during inflation is related to the gravitino mass at present through

$$H_{\text{inf}}^2 \approx \frac{V_*}{3} \sim \frac{|V_{\text{AdS}}|}{3} \approx m_{3/2}^2\,. \tag{8.19}$$

Another mechanism for inflation could be due to the dynamics of branes in the compact space [91]. In this case, the inflaton is some other modulus field σ controlling the location of the D-branes in the internal space. The uplifted KKLT potential then receives a contribution from the inflaton due to the structure of the supergravity F-term potential. The inflationary potential will generically take the form

$$V_{\text{inf}}(\tau) = V_{\text{KKLT}'}(\tau) + \frac{V(\sigma)}{\tau^3}\,. \tag{8.20}$$

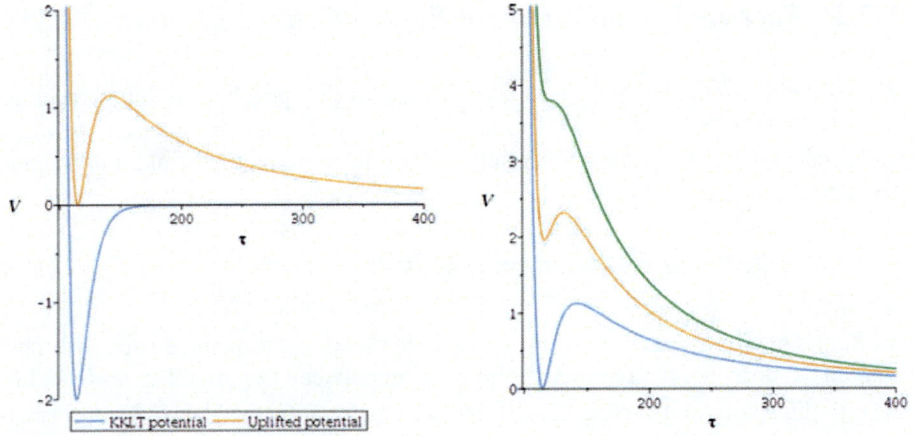

Fig. 8.1 Left: KKLT scalar potential with and without uplift. Right: Uplifted potential with increasing inflationary corrections

As shown in Fig. 8.1, the inflationary corrections raise the level of the minimum more than they raise the height of the barrier. Eventually, if the scale of inflation is high enough, the local minimum turns into an inflection point and the barrier disappears completely. For the volume modulus to stay stabilised during inflation, we require that $V_{\text{inf}} \lesssim V_* \sim 3m_{3/2}^2 M_p^2$, and again we find a relation between the value of the Hubble parameter during inflation and the gravitino mass today

$$H_{\text{inf}}^2 \approx \frac{V_{\text{inf}}}{3M_p^2} \lesssim m_{3/2}^2 \,. \tag{8.21}$$

Equations Eqs. (8.19) and (8.21) are the main results of [59] and show that, unless one can parametrically decouple the value of the gravitino mass from the height of the potential barrier, the scale of inflation is bounded from above by the gravitino mass, $H_{\text{inf}} \lesssim m_{3/2}$. This presents a problem in that it sets the gravitino mass to be extremely large, which by proxy sets the scale of supersymmetry breaking to be much larger than the TeV scale. Notice that this is not necessarily a problem for models with a high scale of supersymmetry breaking. However, to obtain the observed value of the Higgs mass, these require severe fine tuning or a sequestered visible sector on D3 branes at singularities where $M_{\text{soft}} \ll m_{3/2}$ [92]. The tension between low scale supersymmetry and inflation is even more acute in LVS models since the barrier is generically not as high, scaling as $V_* \sim m_{3/2}^3 M_p$. Using similar arguments as above, this constrains the scale of inflation to be $H_{\text{inf}} \lesssim m_{3/2}\sqrt{m_{3/2}/M_p}$.

8.2.2 Racetrack Solution to the KL Problem

Kallosh and Linde [59] proposed a resolution to this problem within a racetrack model [75], see Sect. 7.9 for the details of the model. The key observation of Kallosh and Linde is that the racetrack potential admits a supersymmetric Minkowski vacuum for a critical value of W_0,

$$|W_0|^{\text{crit}} = A\,\mathcal{R}^{\frac{a}{b-a}} + B\,\mathcal{R}^{\frac{b}{b-a}}, \qquad \mathcal{R} = -\frac{aA}{bB}, \tag{8.22}$$

while the gravitino mass vanishes at the Minkowski vacuum since supersymmetry remains unbroken. As such there is no relation between the gravitino mass and the height of the potential barrier. The latter scales as $V_* \lesssim (a-b)^3 A^2 \mathcal{R}^{\frac{2a}{b-a}}$, and can be made arbitrarily high provided we assume $(a-b) \ll 1$, while the minimum for the volume modulus $\tau_{\text{min}} = \frac{1}{a-b}\ln\mathcal{R}$ is pushed to a large value of τ.

8.2.3 General Implications for Quintessence

Although [59] were interested in constraining the scale of inflation and the form of the inflationary potential, similar considerations can be applied to the low energy potential describing the dynamics of dark energy today. We focus on the dynamics of up to three moduli: a quintessence field ϕ describing dark energy, an inflaton σ and the volume modulus τ. In principle, these could be three different moduli, or they could overlap—we consider all possibilities.

We start by assuming they are all different. In general, the full scalar potential can be written as

$$V_{\text{tot}}(\phi,\sigma,\tau) = V_0(\tau) + V_1(\sigma,\tau) + V_2(\phi,\tau) \tag{8.23}$$

where V_0 is the potential that fixes the volume mode, V_1 the contribution of the inflaton and V_2 from quintessence. In this case, dark energy is assumed to be described by a scalar potential $V_{\text{DE}}(\phi,\tau) = V_0(\tau) + V_2(\phi,\tau)$ with ϕ in slow roll at some scale $\phi \sim \phi_0$ today and the volume stabilised at some large value $\tau = \tau_0$. However, during inflation, we generically expect the full scalar potential to receive an inflaton-dependent correction as described in the original KL scenario [59]. In other words,

$$V_{\text{inf}}(\phi,\sigma,\tau) = V_{\text{DE}}(\phi,\tau) + \frac{V(\sigma)}{\tau^3}\,. \tag{8.24}$$

where we have set $V_1(\sigma,\tau) = \frac{V(\sigma)}{\tau^3}$, as a result of Weyl rescaling. The inflaton field rolls slowly through at least 50 efoldings of inflation, starting out at σ_{inf} and ending at σ_0, with $V(\sigma_0) \approx 0$. The volume modulus and the quintessence field are assumed to be spectators during inflation, stabilised at $\tau \approx \tau_{\text{inf}}$ and $\phi \approx \phi_{\text{inf}}$ respectively. Once inflation has ended, the inflaton dumps energy into the Standard Model sector. During this phase of reheating, the volume and quintessence moduli are allowed to move, if necessary, towards their current values, $\tau \to \tau_0, \phi \to \phi_0$. However, in order to avoid potential problems with light element abundances [93] and the spectrum of the cosmic microwave background radiation [2], all three moduli must remain stabilised from nucleosynthesis onwards, right up until the current epoch of dark energy domination, at which point ϕ starts to slow roll.

The key observation is that $V_{\rm DE}(\phi_0, \tau_0) \simeq H_0^2 \ll H_{\rm inf}^2 \simeq V(\sigma_{\rm inf})/\tau_{\rm inf}^3$, where H_0 is the current Hubble scale, and $H_{\rm inf}$ is the scale of inflation. The hierarchy is a considerable one: the scale of dark energy is $H_0 \sim 10^{-60} M_p$, whereas the scale of inflation is assumed to lie somewhere in the range $10^{-42} M_p \lesssim H_{\rm inf} \lesssim 10^{-5} M_p$.[3] The considerations of [59], now suggest a parametric separation in the scale of the underlying quintessence potential in the early universe and at late times. In particular, we can constrain the scale at early times, given by $V_{\rm DE}(\phi_{\rm inf}, \tau_{\rm inf})$ by demanding that the volume modulus remains stabilised during inflation. Following the same logic as [59], we note that to avoid the runaway in the volume, the corresponding minimum at $\tau_{\rm inf}$ should be separated from the asymptotic region by a barrier, V_*, as high as the scale of inflation, $V_* \gtrsim H_{\rm inf}^2$. Given that generically we expect $V_* \sim |V_{\rm DE}(\phi_{\rm inf}, \tau_{\rm inf})|$, to avoid a runaway we require $|V_{\rm DE}(\phi_{\rm inf}, \tau_{\rm inf})| \sim V_* \gtrsim H_{\rm inf}^2 \gg H_0^2 \simeq V_{\rm DE}(\phi_0, \tau_0)$. As we will see in a moment, it is hard to see how we can achieve this separation of scales in a controlled setup.

One of the lessons from Sect. 8.1 is that, at leading order and weak coupling, one scale typically controls the scale of the AdS vacuum and the height of the barrier, both going as $W_0^p/\mathcal{V}^q \ll 1$, for some $p, q > 0$. If the volume, inflaton and quintessence fields, correspond to three different moduli, we have seen how the considerations of Kallosh and Linde [59] suggest that the barrier height should be at least as large as the scale of inflation to avoid a runaway. This fixes the scale of the underlying potential to be far in excess of the dark energy scale, $W_0^p/\mathcal{V}^q \gtrsim H_{\rm inf}^2 \gg H_0^2$. The AdS vacuum, even if it could be uplifted to Minkowksi by the inflationary energy density, would now be too deep for any next to leading order correction to be a viable dark energy candidate, where the potential must be positive.

Although our arguments have focused on the case where the inflation, the volume and quintessence field are three different moduli, the situation is not improved when we relax this assumption. Let us consider each of the alternatives:

- *The volume accounts for dark energy but not for inflation:* The inflaton is once again assumed to be some other (s)axion orthogonal to the volume mode. It is then required that the volume mode stays stabilised during inflation and finds itself in a gentle slope at late times, giving rise to dynamical dark energy. However, in order to avoid the runaway during inflation, the potential must have a large barrier, far in excess of the scale of the late time potential. This scenario is very similar to the one we have already described, and as such, suffers from the same difficulties.

[3] Here, the lower bound comes from the scale of BBN (around MeV), although the actual temperature of the primordial bath might be higher, of $\mathcal{O}$(GeV) [94]. The upper bound comes from constraints on the tensor-to-scalar ratio [2].

Volume driven quintessence will also give rise to long range forces that violate fifth force constraints (see e.g. [55]).

- *The volume accounts for inflation but not for dark energy:* In this scenario, the potential for the volume contains a high scale plateau, allowing the volume to roll slowly during inflation. After inflation, the volume should settle into a low scale Minkowksi vacuum. This could then be stabilised at leading order, with some next-to-leading effect giving rise to dynamical dark energy through another modulus. In [60, 90] volume inflation near an inflection point has been realised by considering different competing contributions: non-perturbative effects, string loops, higher derivative corrections, anti-branes and charged hidden matter fields. Besides looking very contrived and tuned, these constructions raise doubts as to the level of perturbative control since the value of the volume during inflation is relatively small. Moreover, one should make sure that the quintessence field away from the minimum does not result in the volume being destabilised.[4]
- *The volume accounts for everything:* In this case, the volume modulus is responsible for both inflation and dark energy. The situation is similar to the previous case but now we require a flat enough plateau later on as well. This seems to require competing terms at both small volumes (during inflation) and large volumes (during the dark energy period), with a significant hierarchy built in. Such hierarchies would need to be generated by exponentials, which are generated non-perturbatively. This suggests the early time behaviour may not be under perturbative control. Furthermore, if inflation ends with the inflaton rolling in a steep potential, and not approaching a minimum, reheating would need to be non-standard. Crucially the late time behaviour would also fall foul of fifth force constraints.

We can try to get around these problems by assuming that the stabilisation of the volume lies at some low scale, near Minkowski vacuum generated at leading order, breaking the connection between the scale of the vacuum and the height of the barrier. (Recall that the barrier height should exceed the scale of inflation to avoid decompactification.) If this leading order stabilisation leaves, say, an axionic flat direction which is lifted only at subdominant order by tiny non-perturbative effects, one could reproduce the required hierarchy between $H_{\rm inf}$ and H_0 without inducing any destabilisation of the volume mode. Notice, however, that generating a supersymmetric Minkowski minimum ($W = 0$) by solving the F-terms equations ($D_i W = 0$) requires a finely tuned cancellation between all contributions to the superpotential, both at tree and non-perturbative level, as in the racetrack scenario [75], which was already identified as a way to skirt around the original KL problem [59]. Moreover, axion quintessence in agreement with swampland bounds on the associated decay constant, requires dynamical dark energy to occur close to the maximum of the axion potential where the scale of supersymmetry breaking would

[4] Notice that the tension between $H_{\rm inf}$ and H_0 could be relaxed by also having W_0 evolve from large to small values during inflation, as in the toy model of [95]. However we are not aware of a robust model that realises this effect while remaining under computational control.

be extremely low, set by the scale of dark energy. This leads us to conclude that *non*-supersymmetric Minkowski vacua are actually more appealing, at least if we want to build a viable model of quintessence in string theory.

8.2.4 A Closer Look at Quintessence Models with a KL Problem

When we consider quintessence in string theory, commitment to the refined dS and weak gravity conjectures forbids a dynamical model of dark energy based on either of the following scenarios:

1. Minkowski vacuum with saxion slow roll down a very shallow potential [61]
2. Minkowski vacuum with axion quintessence with trans-Planckian decay constant [71].

Whilst this leaves some alternatives, our consideration of the KL problem in the previous section suggests that most of these are also ruled out. In particular, the following scenarios

1. Saxion hilltop for a Minkowski or AdS vacuum
2. Axion hilltop for a Minkowski vacuum with no hierarchy
3. Saxion slow roll down a moderate slope, with a runaway or a Minkowski vacuum.

are all compatible with the dS conjecture. Two minor clarifications are in order here. By 'hierarchy' we mean the existence of an exponential hierarchy of scales between the leading order potential for the volume and the axion potential responsible for quintessence. By a 'moderate slope' we mean order one in Planck units, i.e. steep enough to satisfy the refined dS conjecture but shallow enough to allow for at least one efolding of slow roll.

Each of these three alternatives suffers from the KL problem. They also suffer from a variety of other problems, not least that of an unacceptably light volume modulus and a light gravitino. In this section we study specific examples of each scenario, explicitly demonstrating how many of these problems emerge.

Saxion hilltop for a Minkowski or AdS vacuum

The racetrack scenario has a supersymmetric Minkowski vacuum, separated from the runaway regime by a maximum in the volume mode. We can therefore imagine a dynamical model of dark energy where the volume mode is rolling close to the hilltop, and the axion is fixed at its minimum, at $\theta = 0$. As we saw previously, the racetrack scenario was proposed as a way around the original KL problem, since the height of the barrier can be taken to be higher than the scale of inflation without any consequence on the gravitino mass. However, the height of the barrier is the height of the maximum in the volume direction (or better, the height of saddle in the (τ, θ)-plane). For hilltop quintessence driven by the volume mode, this height is now set by the dark energy scale

$$H_0^2 \sim V_{\rm race}(\tau_{\rm max}, 0). \tag{8.25}$$

Clearly this barrier is too small to protects us from the KL problem. Indeed, the contribution from inflation, driven by a different field σ, couples to the volume mode due to Weyl rescaling. As expected, it will induce destabilisation of the volume towards decompactification since

$$\frac{V(\sigma)}{\tau^3} \sim H_{\rm inf}^2 \gg H_0^2 \sim V_{\rm race}(\tau_{\rm max}, 0) \tag{8.26}$$

Of course, similar considerations also apply to saxion hilltops where the global minimum is supersymmetric AdS. As an example, consider KKLT models where a hilltop in the volume modulus is generically present, even in the absence of an anti-D3 brane uplift, as a consequence of α'^3 corrections to the Kähler potential, as already stressed in [6]. This model is also tractable enough to easily demonstrate other issues that can emerge beyond the KL problem, such as the light volume modulus and the light gravitino. Let us run through some of the details.

The KKLT potential was already given in Eq. (8.15). If we assume that the axion θ is stabilised at its minimum at $\theta = 0$, the dynamics of the volume modulus τ in a neighbourhood of the maximum is controlled by the last two terms, in other words

$$V_{\rm KKLT}(\tau, 0) \approx -\frac{aA|W_0|\,e^{-a\tau}}{2\tau^2} + \frac{3W_0^2\hat{\xi}}{64\sqrt{2}\tau^{9/2}}, \tag{8.27}$$

with $|W_0|$ given by Eq. (7.48). This simplification allows us to show that the maximum is located at $\tau_{\rm max}$, defined by the relation

$$\hat{\xi} = \frac{64\sqrt{2}aA\tau_{\rm max}^{5/2}e^{-a\tau_{\rm max}}(a\tau_{\rm max}+2)}{27|W_0|}. \tag{8.28}$$

Since $a\tau_{\rm max} \gg 1$, the height of the potential at the maximum is

$$V_{\rm KKLT}(\tau_{\rm max}, 0) \approx \frac{a^2A|W_0|\,e^{-a\tau_{\rm max}}}{9\tau_{\rm max}} = \frac{4}{9}a\tau_{\rm min}\left(\frac{\tau_{\rm min}}{\tau_{\rm max}}\right)e^{-a\Delta\tau}|V_{\rm AdS}| \tag{8.29}$$

where we made use of Eq. (7.48) and the expression for the scale of the leading order AdS minimum Eq. (8.16). The minimum at $\tau_{\rm min}$ and the maximum at $\tau_{\rm max}$ are separated by distance $\Delta\tau = \tau_{\rm max} - \tau_{\rm min} > 0$.

Clearly the height of the maximium should be fixed by the current Hubble scale, $H_0^2 \sim V_{\rm KKLT}(\tau_{\rm max}, 0)$. By the same reasoning as for the racetrack scenario, we run into a KL problem. In this simple model, it is also instructive to demonstrate the smallness of the mass of the gravitino and the volume modulus explicitly.

Current observational bounds require $V_{\rm KKLT}(\tau_{\rm max}, 0) \sim 10^{-120}$ in Planck units. Such low values can be achieved either by having $\tau_{\rm max}$ large or $|W_0|$ exponentially small, two requirements that are not independent in KKLT, as can be seen from

Eq. (7.48). In order to estimate the choice of parameters that leads to the correct value for H_0, we use the fact that the maximum and the minimum are not too far apart, and compute the height of the maximum to zeroth order in $\Delta\tau$

$$V_{\text{KKLT}}(\tau_{\text{max}}, 0) \approx \frac{4}{9} a\tau_{\text{min}} |V_{\text{AdS}}| \approx \frac{2a^3 A^2}{27} e^{-2a\tau_{\text{min}}} \tag{8.30}$$

where we made use of Eq. (8.16). Assuming $\frac{2a^3 A^2}{27} = \mathcal{O}(1)$, matching the observed value of H_0 requires $a\tau_{\text{min}} \sim 140$, which through Eq. (7.48) translates as $|W_0| \sim 10^{-59}$ and a gravitino mass of $m_{3/2} \sim 10^{-33}$ eV. This is unacceptably light [64, 67]. The fact that the gravitino mass is of order the dark energy scale can be traced back to the fact that the leading order vacuum is supersymmetric and very close to Minkowski. Indeed, from Eq. (8.29), we see that the scale of the supersymmetric AdS vacuum is bounded above by the dark energy scale. This failure to decouple $m_{3/2}$ and H_0 is clearly typical of any model featuring a leading order supersymmetric Minkowski, or near Minkowski, vacuum.

To compute the mass of the volume mode, one has first to switch to a canonical field via $\phi = \sqrt{\frac{3}{2}} \ln \tau$ and then compute $m_\phi^2 \simeq V_{\phi\phi}$ at the location of the maximum. This yields $m_\phi^2 \simeq -3a\tau_{\text{max}} V_0 \simeq -3a\tau_{\text{max}} m_{3/2}^2$. Since $a\tau_{\text{max}} \gtrsim a\tau_{\text{min}} \sim 140$, this implies that the mass of the volume mode is only one order of magnitude above the gravitino mass, explicitly showing the existence of a light volume problem.

Axion hilltop for a Minkowski vacuum with no hierarchy

Let us return to the racetrack model and consider using θ, instead of τ, to drive quintessence. Once again, since the dark energy scale now sets the scale of the potential; this will immediately run into a KL problem. As it happens, this model suffers from another problem, closely related to the KL problem, but applied only to late time dynamics. Indeed, even if we ignore the contributions from inflation, the volume barrier disappears as soon as we move the axion sufficiently far away from its minimum. In other words, in attempting to move the axion to the hilltop, the volume itself is immediately destabilised.

This is demonstrated numerically in Fig. 8.2. Here we plot the form of the racetrack potential Eq. (7.53) as a function of the volume modulus τ for different values of the axion θ. When the axion lies at its minimum at $\theta = 0$, we see that the volume is stabilised at the Minkowski minimum. However, as we increase θ in units of the instanton coupling a, the volume barrier begins to shrink, and eventually disappears completely. At this point the volume will roll towards the AdS vacuum and any hope of exploiting the axion as a dynamical dark energy model is lost.

These problems might have been anticipated in the racetrack scenario, as both the stabilisation of the volume and the dynamics of the corresponding axion rely on the *same* non-perturbative corrections to the superpotential. There was always a danger that the stabilisation would fail the moment the axion began to roll. As already pointed out, to proceed with a viable model of quintessence, we need to break the connection between the stabilisation of the volume and the dynamics of the would-be dark energy field, creating a hierarchy in mass between these two fields.

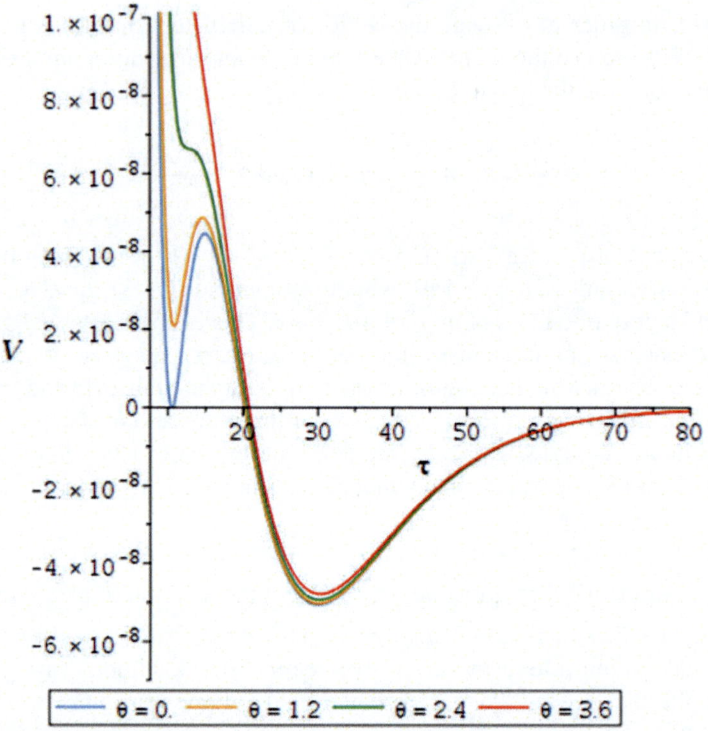

Fig. 8.2 Racetrack potential at different values of the axion θ for parameter choice $A = 1, B = -1, a = 0.1, b = 0.09$

Saxion slow roll down a moderate slope, with a runaway or a Minkowski vacuum

Let us now focus on a saxion runaway model, where the saxion is asymptotically rolling slowly down a moderate slope. At leading order, our example contains a non-supersymmetric Minkowski vacuum where one of the saxion directions is flat. The saxion runaway potential is then generated perturbatively, beyond leading order. However, since it is perturbative, it is not possible to generate a large enough hierarchy between the leading and subleading order terms to prevent the KL problem and destabilising the volume. Note that similar considerations would apply if additional subleading corrections were to generate a global non-supersymmetric Minkowski minimum, as opposed to a runaway. As shown in [96], the case with a global supersymmetric Minkowski minimum is actually incompatible with slow roll down a moderate slope due to the stability condition on the form of the scalar potential.

Consider a fibred Calabi-Yau whose volume takes the form [97, 98]

$$\mathcal{V} = \sqrt{\tau_1}\tau_2 \tag{8.31}$$

The saxion kinetic terms look like (we ignore the corresponding axions)

$$\mathcal{L}_{\text{kin}} = \frac{1}{2}\left[(\partial \ln \tau_2)^2 + \frac{1}{2}(\partial \ln \tau_1)^2\right] \tag{8.32}$$

and can be brought into canonical form by the following field redefinition

$$\tau_1 = e^{\sqrt{\frac{2}{3}}\chi + \frac{2}{\sqrt{3}}\phi} \qquad \tau_2 = e^{\sqrt{\frac{2}{3}}\chi - \frac{1}{\sqrt{3}}\phi} \tag{8.33}$$

Notice that χ corresponds to the volume mode $\mathcal{V}$, and ϕ to the ratio $u = \tau_1/\tau_2$ since

$$\mathcal{V} = \sqrt{\tau_1}\tau_2 = e^{\sqrt{\frac{3}{2}}\chi} \qquad u = \frac{\tau_1}{\tau_2} = e^{\sqrt{3}\phi} \tag{8.34}$$

Let us consider an effective field theory defined by the following Kähler potential and superpotential

$$K = -2\ln\left(\mathcal{V} + \frac{\xi}{2g_s^{3/2}} - \gamma\sqrt{g_s}\ln\mathcal{V}\right) - \frac{\tilde{C}}{\mathcal{V}\sqrt{\tau_1}} \qquad W = W_0 \tag{8.35}$$

where ξ controls $O(\alpha'^3)$ corrections, while γ controls brane loop corrections at $O(\alpha'^3 g_s^2)$ [29]. For $\mathcal{V} \gg 1$ these can naturally compete with the tree-level $O(\alpha'^3)$ term due to the $\ln\mathcal{V}$ enhancement factor. The term proportional to $\tilde{C}$ represents $O(\alpha'^4 g_s^2)$ string loop corrections due to exchange of winding modes at the intersection of D7-branes. This contribution is subleading since it is suppressed by an additional power of α' with respect to the terms proportional to ξ and γ. We do not include Kaluza-Klein loop corrections since they would be suppressed with respect to winding contributions by an additional power of g_s^2, and, moreover, they could be absent by construction if all branes intersect each other. We also neglect higher derivative F^4 contributions to the scalar potential since they would arise with additional volume suppression factors.

The Kähler potential and superpotential in (8.35) generate the following scalar potential

$$V = V_{\text{lead}}(\mathcal{V}) + V_{\text{sub}}(\mathcal{V}, u) \tag{8.36}$$

where

$$V_{\text{lead}}(\mathcal{V}) = \frac{C_{\text{up}}}{\mathcal{V}^{8/3}} + \frac{3W_0^2}{4\mathcal{V}^3}\left(-2\gamma\sqrt{g_s}\ln\mathcal{V} + \frac{\xi}{g_s^{3/2}}\right) \tag{8.37}$$

and

$$V_{\text{sub}}(\mathcal{V}, u) = \frac{2\tilde{C}W_0^2}{\mathcal{V}^{10/3}}\frac{1}{u^{1/3}} \tag{8.38}$$

Notice that in V_{lead} we included also a term proportional to C_{up} representing the positive contribution of a T-brane background [99] which is a generic feature of type

IIB compactifications with 3-form fluxes and magnetised D7-branes. In the limit where the supergravity approximation is under control, i.e. for $\mathcal{V} \gg 1$ and $g_s \ll 1$, $V_{\rm sub}$ is indeed subdominant with respect to $V_{\rm lead}$ since

$$\frac{V_{\rm sub}}{V_{\rm lead}} \sim \frac{g_s^{3/2}}{\mathcal{V}^{1/3}} = \frac{g_s^2}{\mathcal{V}_s^{1/3}} \ll 1 \tag{8.39}$$

where $\mathcal{V}_s = \mathcal{V} g_s^{3/2}$ denotes the string frame volume. Thus at leading order the potential features a flat direction parametrised by u. At this level of approximation, by a suitable tuning of $C_{\rm up}$, $V_{\rm lead}$ features a non-supersymmetric Minkowski minimum where the volume scales as (for $k = \xi/\gamma$)

$$\mathcal{V}_{\rm min} \sim e^{\frac{k}{g_s^2}} \gg 1 \qquad \text{for} \qquad g_s \ll 1 \tag{8.40}$$

The subleading contribution $V_{\rm sub}$ determines just a small shift of the volume minimum and generates a runaway for u which, when written in terms of the canonically normalised field ϕ, looks like

$$V_{\rm sub}(\phi) = V_0\, e^{-\lambda\phi} \quad \text{with} \quad \lambda = \frac{1}{\sqrt{3}} \quad \text{and} \quad V_0 = \frac{2\tilde{C} W_0^2}{\mathcal{V}_{\rm min}^{10/3}} \tag{8.41}$$

This gentle runaway could provide an interesting model of quintessence in agreement with the refined dS conjecture since $\lambda \simeq 0.577$ is of order unity and it marginally satisfies the bound $\lambda \leq 0.6$ obtained in [63]. However the requirement to avoid volume destabilisation due to the inflationary energy implies (similar considerations would apply also to the case where (8.41) describes a quintessence potential with a global Minkowski minimum)

$$V_{\rm lead} = \left(\frac{V_{\rm lead}}{V_{\rm sub}}\right) V_{\rm sub} \sim \frac{\mathcal{V}^{1/3}}{g_s^{3/2}} H_0^2 \gtrsim H_{\rm inf}^2 \quad \Leftrightarrow \quad \frac{\mathcal{V}^{1/3}}{g_s^{3/2}} \gtrsim \left(\frac{H_{\rm inf}}{H_0}\right)^2 \gtrsim 10^{-36}, \tag{8.42}$$

where we have used $H_{\rm inf} \gtrsim 10^{-42}\, M_p$ as the extreme lower bound on the Hubble scale during inflation to be compatible with a BBN reheating temperature of at least 1 MeV. Using (8.40) which implies $g_s \sim (\ln \mathcal{V})^{-1/2}$, this bound becomes

$$\mathcal{V}^{1/3} (\ln \mathcal{V})^{3/4} \gtrsim 10^{36} \quad \Leftrightarrow \quad \mathcal{V} \gtrsim 10^{103} \tag{8.43}$$

This would yield a string scale M_s well below the TeV scale and a gravitino mass $m_{3/2}$ well below the meV scale since

$$M_s \simeq g_s^{1/4} \frac{M_p}{\sqrt{\mathcal{V}}} \sim \frac{M_p}{(\ln \mathcal{V})^{1/8} \sqrt{\mathcal{V}}} \lesssim 10^{-52}\, M_p \tag{8.44}$$

$$m_{3/2} \simeq g_s^{1/2} \frac{M_p}{\mathcal{V}} \sim \frac{M_p}{(\ln \mathcal{V})^{1/4}\, \mathcal{V}} \lesssim 10^{-104}\, M_p \tag{8.45}$$

Similar considerations imply that the mass of the volume mode is also very suppressed with respect to the meV scale. Hence the hierarchy between $V_{\rm lead}$ and $V_{\rm sup}$ is not big enough to prevent the KL and light volume problems. The reason is that the effective shift symmetry for u is already broken at perturbative level.

8.3 Axion Hilltop Quintessence and Initial Conditions

As explained in the previous section, a viable quintessence model has to feature a leading order non-supersymmetric Minkowski vacuum with hierarchy, i.e. where, at leading order, the axion is a flat direction while the saxion (in particular, the volume mode) is heavy. The axionic flat direction is then lifted by subdominant instanton effects which can lead to axion hilltop quintessence. In this section we therefore focus on this model, providing first an explicit realisation in LVS string models, and then studying the issue of initial conditions.

8.3.1 LVS Axion Hilltop Quintessence

The simplest way to realise an axion hilltop quintessence model in type IIB string theory is through the simplest LVS scenario, with two Kähler moduli $T_b = \tau_b + {\rm i}\theta_b$ and $T_s = \tau_s + {\rm i}\theta_s$, where the 'big' modulus τ_b turns out to be much larger than the 'small' modulus τ_s. The important point is that the scalar potential only depends on the volume axion, θ_b, at next to leading order, without affecting the stabilisation of the volume mode which occurs at leading order. As a result, the volume axion can potentially play the rôle of quintessence when it is rolling near the top of its potential, without having any of the adverse consequences we saw for the racetrack scenario. We should, however, be mindful of the fact that the simplest LVS setup leads to a non-supersymmetric AdS vacuum which needs to be uplifted to Minkowski by the inclusion of additional sources of energy like T-branes [99]. Notice that these positive contributions to the scalar potential are generic features of consistent type IIB compactifications [10, 12, 100–102] due to the presence of hidden sector stacks of D7-branes (induced by D7 tadpole cancellation), 2-form gauge fluxes (induced by Freed-Witten anomaly cancellation) and 3-form background fluxes (used to freeze the dilaton and the complex structure moduli). For further details and a comprehensive discussion of dS model building in string theory, see [53, 103].

The main ingredients of this model are $\mathcal{O}(\alpha'^3)$ contributions to the Kähler potential and non-perturbative corrections to the superpotential of the form

$$K = -2\ln\left(\mathcal{V} + \frac{\hat{\xi}}{2}\right) \qquad W = W_0 + A_s\, e^{-a_s T_s} + A_b\, e^{-a_b T_b} \tag{8.46}$$

where the internal geometry corresponds to a simple Swiss-cheese scenario with the volume given by $\mathcal{V} = \tau_b^{3/2} - \tau_s^{3/2}$. The resulting potential looks like

$$V = \frac{4A_b^2 a_b^2}{3\tau_b} e^{-2a_b\tau_b} + \frac{8A_s^2 a_s^2 \sqrt{\tau_s}}{3\tau_b^{3/2}} e^{-2a_s\tau_s} + \frac{3|W_0|^2\hat{\xi}}{4\tau_b^{9/2}} - \frac{4A_b a_b}{\tau_b^2}|W_0|\, e^{-a_b\tau_b}\cos(a_b\theta_b)$$
$$- \frac{4A_s a_s \tau_s}{\tau_b^3}|W_0|\, e^{-a_s\tau_s}\cos(a_s\theta_s) + \frac{8A_b A_s a_b a_s \tau_s}{\tau_b^2} e^{-(a_b\tau_b + a_s\tau_s)}\cos(a_b\theta_b - a_s\theta_s)\,,$$

where we have used $\tau_b \gg \tau_s$, $a_b\tau_b \gg 1$, $a_s\tau_s \gg 1$ and we have assumed $W_0 < 0$ so that the axions are minimised at $\theta_b = \theta_s = 0$. With the axions settled at their minima, we consider the stabilisation of τ_s and τ_b. Dropping all terms suppressed by $e^{-a_b\tau_b}$ or more, one finds the well-known LVS results from the variation with respect to τ_s and τ_b

$$\mathcal{V} \simeq \langle\tau_b\rangle^{3/2} \simeq \frac{3|W_0|\sqrt{\langle\tau_s\rangle}}{4A_s a_s} e^{a_s\langle\tau_s\rangle} \qquad \frac{\hat{\xi}}{2} \simeq \langle\tau_s\rangle^{3/2} \simeq \left[\frac{1}{a_s}\ln\left(\frac{\mathcal{V}}{|W_0|}\right)\right]^{3/2} \tag{8.47}$$

The minimum is AdS, breaks supersymmetry and, to leading order, is given by

$$\langle V\rangle \simeq -\frac{3\hat{\xi}|W_0|^2}{8a_s\langle\tau_s\rangle\mathcal{V}^3}. \tag{8.48}$$

There are several sources of uplifting to Minkowski which can be expressed as

$$V_{\text{up}} = \frac{\kappa}{\mathcal{V}^\alpha}, \tag{8.49}$$

where κ is a positive coefficient and $0 < \alpha < 3$. For example, $\alpha = 8/3$ for T-branes while $\alpha = 4/3$ for an anti D3-brane at the tip of a warped throat, although the particulars of the uplifting mechanism are unimportant for the discussion that follows. The uplift term modifies the second relation in Eq. (8.47) as

$$\frac{\hat{\xi}}{2} = \langle\tau_s\rangle^{3/2} - \frac{2\alpha\kappa}{9|W_0|^2}\mathcal{V}^{3-\alpha}, \tag{8.50}$$

and we will fix κ, to zeroth order in $e^{-a_b\tau_b}$, by demanding that the uplifted LVS vacuum is Minkowski, i.e.

$$\langle V\rangle = -\left(\frac{|W_0|^2\langle\tau_s\rangle^{3/2}}{\mathcal{V}^3} - \frac{\kappa\,(3-\alpha)}{3\mathcal{V}^\alpha}\right) = 0$$
$$\Rightarrow \quad \kappa = \frac{3|W_0|^2\langle\tau_s\rangle^{3/2}}{(3-\alpha)\,\mathcal{V}^{3-\alpha}}. \tag{8.51}$$

The Eqs. (8.47), (8.50) and (8.51) form a system fixing $(\langle\tau_b\rangle, \langle\tau_s\rangle, \kappa)$ for a particular choice of $(\xi, W_0, A_s, A_b, a_s, a_b, \alpha)$. Once the Minkowski vacuum is fixed in this way, we focus on θ_b as a dark energy candidate. The hierarchy of scales between $e^{a_b\tau_b}$ and $e^{a_s\tau_s}$ guarantees that shifts in the θ_b direction do not destabilise the Minkowski vacuum.

The uplift term can be further adjusted at $\mathcal{O}\left(e^{-a_b\tau_b}\right)$ to guarantee a Minkowski vacuum at $\tau_b = \langle\tau_b\rangle$, $\tau_s = \langle\tau_s\rangle$ and $\theta_b = \theta_s = 0$. Releasing the volume axion, θ_b, its dynamics is then described by the following dark energy potential to leading order

$$V_{\rm DE} = \left[\frac{4A_b a_b}{\langle\tau_b\rangle^2}|W_0|\, e^{-a_b\langle\tau_b\rangle} - \frac{8A_bA_s a_b a_s\langle\tau_s\rangle}{\langle\tau_b\rangle^2}\, e^{-(a\langle\tau_b\rangle + a_s\langle\tau_s\rangle)}\right](1-\cos(a_b\theta_b))$$
$$\simeq \frac{4A_b a_b}{\langle\tau_b\rangle^2}|W_0|\, e^{-a_b\langle\tau_b\rangle}\,(1-\cos(a_b\theta_b))\,, \tag{8.52}$$

where we explicitly see the Minkowski minimum at $\theta_b = 0$. The maximum is located at $\theta_b = \pi/a$. From the form of the Kähler metric, the canonically normalised axion and the corresponding decay constant turn out to be

$$\phi \simeq \sqrt{\frac{3}{2}}\frac{\theta_b}{\langle\tau_b\rangle} \qquad f_a = \sqrt{\frac{3}{2}}\frac{M_P}{a_b\langle\tau_b\rangle}, \tag{8.53}$$

so that (8.52) can be rewritten in a more standard way as

$$V_{\rm DE} = V_0\left(1-\cos\frac{\phi}{f_a}\right) \qquad \text{where} \qquad V_0 \equiv \frac{4A_b a_b}{\langle\tau_b\rangle^2}|W_0|\, e^{-a_b\langle\tau_b\rangle} \tag{8.54}$$

The η parameter at the maximum of the dark energy potential, where $\theta_b = \pi/a_b$, becomes

$$\eta_{\rm hilltop} = \left.\frac{V_{{\rm DE},\,\phi\phi}}{V_{\rm DE}}\right|_{\phi_{\rm max}} = \frac{2}{3}\langle\tau_b\rangle^2 \left.\frac{V_{{\rm DE},\,\theta_b\theta_b}}{V_{\rm DE}}\right|_{\theta_b=\pi/a_b} = -\frac{1}{3}a_b^2\langle\tau_b\rangle^2 \tag{8.55}$$

To estimate this, notice that the value of the potential at the hilltop should be $\mathcal{O}(10^{-120})$ in Planck units to be compatible with dark energy at late times. This suggests $a_b\langle\tau_b\rangle \sim \mathcal{O}(100)$, and so $\eta_{\rm hilltop} \sim -\mathcal{O}(3000)$. Clearly the curvature at the hilltop is compatible with the improved swampland bound of Eq. (8.2). However, the large absolute value of the η parameter requires a high degree of fine tuning of the initial conditions for θ_b if it is to give rise to a viable quintessence model as it rolls away from the maximum, as we will illustrate in Sect. 8.3.2.

Other approaches to axion hilltop quintessence

Before discussing the issue of initial conditions, let us mention other two possible approaches which can lead to a viable quintessence models via axion hilltop:

1. *Uplifted KKLT with an orientifold-odd axion*
   The standard uplifted KKLT scenario with a single Kähler modulus $T = \tau + i\theta$

features a non-supersymmetric Minkowski vacuum with no flat direction. This cannot be used to drive quintessence since both τ and θ are lifted by the same non-perturbative effect. However, in the presence of an extra orientifold-odd modulus $G = c + Sb$ (where S is the axio-dilaton), b would also be lifted by the non-perturbative superpotential e^{-aT} (with a minimum at $b = 0$), while the axionic mode c would remain flat. This axionic direction can, instead, be lifted at subleading order by the inclusion of fluxed E3-instanton corrections to W of the form $e^{-a(T+S+iG)} \sim e^{-aT} e^{-a/g_s}$ for $b = 0$ [104]. Therefore, for $g_s \ll 1$, the scale of the potential for c is exponentially suppressed with respect to the potential for T, providing a promising candidate for a viable quintessence model with decay constant $f_a \simeq \sqrt{g_s/\tau}\, M_p \ll M_p$ for $\tau \gg 1$ [105].

2. *Non-geometric fluxes*

A second possibility is to consider the effect of non-geometric fluxes which extend the GVW superpotential to [13, 106–108]

$$W = P_1(U) + S P_2(U) + \sum_i T_i P_3^{(i)}(U), \tag{8.56}$$

where $P_3^{(i)}$ are cubic polynomials of the complex structure sector U. Combined with the tree-level expression of the Kähler potential, the dependence of W on T generates no-scale breaking contributions to the scalar potential. With regards to the previous discussion, if we are able to stabilise all but one (axionic) modulus at a non-supersymmetric Minkowski vacuum (or even a near Minkowski AdS minimum) at tree-level through an appropriate choice of fluxes, then the hierarchy between non-perturbative effects and tree-level would guarantee that lifting the leftover flat direction would not displace the heavy moduli from the tree-level minimum. Furthermore, the leftover axion could be made parametrically light and may be used to drive quintessence.[5]

8.3.2 Constraining Hilltops

Hilltop models are classically unstable. The rate of the instability is controlled by the η parameter, describing the rate of change of the gradient close to the maximum of the potential. The larger the η parameter, in absolute value, the closer the field needs to start near the maximum in order to obtain the required period of acceleration. Fortunately, for quintessence, we only require one efolding of accelerated expansion (this is in contrast to early universe inflation which requires at least 50). Nevertheless, in string theory, the η parameter can sometimes be quite large forcing the field to start very close to the top of the hill. Classically, this is not problematic if one accepts the inevitable tuning of initial conditions, although as we will see later, quantum

[5] Assuming that backreaction effects on the Kähler potential can be kept under control and the internal volume can be made large enough to trust the perturbative expansion.

diffusion at early times can push the field away from the sweet spot, spoiling any realistic chance of late time quintessence.

In this section we will derive the constraints on the parameters and the initial conditions of a generic model of late time acceleration. As explained in the previous sections, our main interest will be in axion hilltop models, although we will also generalise our analysis also to saxion hilltop models which we approximate in a neighbourhood of the maximum as an inverted quadratic.

Axion hilltop quintessence

In the context of late time acceleration, axions are the prototype of thawing quintessence models [109, 110]: models where the field is frozen due to Hubble friction until the very recent past. These models are known to be sensitive to the choice of initial conditions and relatively insensitive to the particular form of the potential. A generic axion potential has the usual trigonometric form

$$V = V_0 \left(1 - \cos \frac{\phi}{f_a}\right) \tag{8.57}$$

where we have (for simplicity) assumed that the vacuum expectation value of the axion lies at vanishing potential, consistent with a Minkowski vacuum. This can lead to accelerated expansion in two distinct regimes $f_a > M_p$ and $f_a < M_p$. For $f_a > M_p$ acceleration takes place in the concave region of the potential, whereas if $f_a < M_p$ it happens in the vicinity of the maximum. While models with super-Planckian decay constants are less sensitive to initial conditions, getting these large values for f_a has proven challenging from a UV point of view due to the tension with the weak gravity conjecture [71] and with explicit computations [70, 105]. For example, the LVS axion model presented in Sect. 8.3.1 features for example a sub-Planckian decay constant since (8.53) gives $f_a \simeq M_p/(a_b\langle\tau_b\rangle) \sim 0.01\, M_p$.

Of course, axion hilltop quintessence can take place irrespective of the value of the decay constant, though it may require finely tuned initial conditions. In Fig. 8.3 we plot the deviation from the maximum as a function of the decay constant for a range of f_a that is compatible with swampland constraints. We see that the range of f_a that is more naturally achieved in UV constructions, $f_a < M_{\rm GUT}$, is also the one that suffers from an extreme sensitivity to the initial position of the field. In the region $f_a \in [0.02, 0.1]\, M_p$, the curve bounding the viable region can be approximated by

$$\ln \Delta_{\rm max} = c_0 + c_1 \ln f_a + c_2 (\ln f_a)^2, \tag{8.58}$$

where $\Delta_{\rm max}$ denotes the maximum distance from the maximum compatible with late time acceleration, $c_0 = -32.6$, $c_1 = -28.977$ and $c_2 = -8.2302$.

Saxion hilltop quintessence

For completeness, we now turn our attention to saxion models of quintessence, which, in the vicinity of the hilltop can approximated by an inverted quadratic

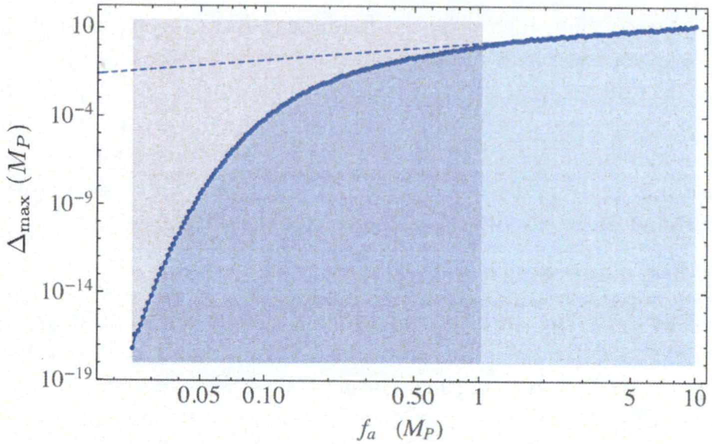

Fig. 8.3 Constraints on initial axion displacement from the maximum compatible with a viable quintessence hilltop model as a function of the decay constant f_a. For at least one e-fold of accelerated expansion, the initial value ϕ_{in} should satisfy $|\phi_{\text{in}} - \phi_{\text{max}}| < \Delta_{\text{max}}$, where ϕ_{max} is the location of the maximum and Δ_{max} is given by the solid blue line. The dashed line shows the position of the inflection point $\phi = \pi f_a/2$. The blue shaded region corresponds to $f_a > M_p$ which is in tension with the weak gravity conjecture

$$V = V_0 - \frac{1}{2} m^2 \phi^2. \tag{8.59}$$

It is useful to define the following parameter

$$\eta_0 = \left. \frac{V_{\phi\phi} M_p^2}{V} \right|_{\phi=0} = -\frac{(m M_p)^2}{V_0} \tag{8.60}$$

which describes the curvature of the scalar potential at the origin. Let us recall that the swampland conjecture (8.2) requires $|\eta_0| \le c' \sim \mathcal{O}(1)$.

Hilltop models, no matter how steep the potential (or how large η_0), *classically* always lead to extended periods of accelerated expansion, given that at the maximum $\epsilon_V = 0$. This would allow for a description of the late time acceleration, regardless of the swampland limits on the slope and curvature of the potential that can be attained within a UV complete framework. On the flip-side, this comes at the price of tuning the initial position of the quintessence field, $|\phi_{\text{in}}| < \Delta_{\text{max}}$—the steeper V, the closer ϕ_{in} needs to be to the maximum in order to have an extended period of acceleration. One would think that this problem is slightly less of an issue in quintessence models which require only $\mathcal{O}(1)$, instead of $\mathcal{O}(50)$, efoldings as in inflation. However for large η_0 the level of tuning is similar.

In Fig. 8.4 we show the maximal allowed initial displacement from the maximum, Δ_{max}, compatible with late time acceleration as a function of η_0 for quadratic quintessence models. Each point corresponds to numerical solutions that start in matter domination with zero initial velocity and different values for ϕ_{in}. We see that

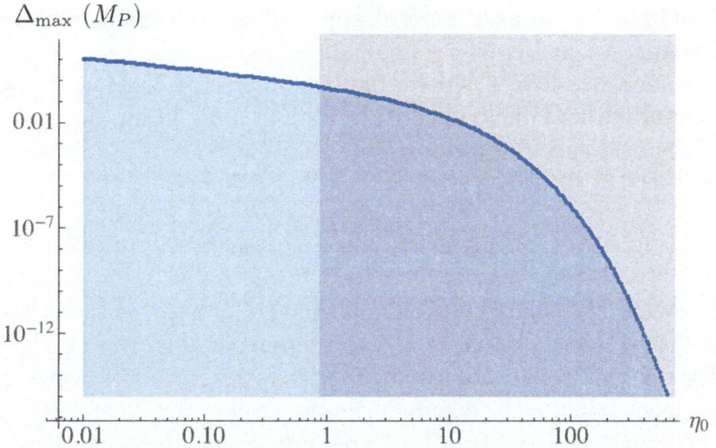

Fig. 8.4 Constraints on initial saxion displacement from the maximum compatible with a viable quintessence hilltop model as a function of η_0. The grey region corresponds to $\eta_0 > 1$ which is in tension with the refined swampland conjecture. The steeper the maximum (larger η_0), the more fine tuned the initial conditions (smaller $\Delta_{\max}$)

the steeper the potential the more tuned is the initial value of ϕ. For $\eta_0 \geq 1$ (grey region) ϕ_{in} must lie within a fraction of M_p from the top of the hill. In the absence of a dynamical mechanism, such initial conditions look rather unnatural. The curve bounding the viable region can be approximated by

$$\ln \Delta_{\max} = c_1 + c_2\, \eta_0^{-p}, \tag{8.61}$$

where $c_1 = 1.7$, $c_2 = -2.1$ and $p = 0.44$.

Even if one is willing to accept this level of tuning in order to describe the observed accelerated expansion and be in agreement with putative bounds from a UV theory, one must ponder if quantum effects will spoil the required tuning of the initial conditions. We address this issue in Sect. 8.3.3.

8.3.3 *Stochastic Effects and Initial Conditions*

In this section we investigate whether the judicious choices of initial conditions described in the previous section survive the unavoidable stochastic fluctuations in the early universe. For our purposes we model inflation as an exact dS background, and therefore fix H_{inf} to be constant. Generally speaking this approximation is adequate when H_{inf} varies slowly, like in plateau models, but needs to be refined in the context of monomial inflationary models as shown in [111]. In the cases of interest, where the stochastic processes are diffusion dominated and the equilibrium distribution (if it exists) is of little relevance, the exact dS approximation gives an adequate description of the system.

The quintessence field ϕ, due to the large hierarchy $V(\phi) \ll H_{\text{inf}}^2 M_p^2$ behaves as a spectator during inflation and is, to leading order in the slow roll expansion, described by the Langevin equation

$$\frac{\partial \phi}{\partial N} = -\frac{V_\phi}{3H_{\text{inf}}^2} + \frac{H_{\text{inf}}}{2\pi}\xi, \tag{8.62}$$

where N denotes the number of efoldings and $\xi(N)$ is a stochastic variable with unit variance $\langle \xi(N_1)\xi(N_2)\rangle = \delta(N_1 - N_2)$ and zero mean $\langle \xi(N)\rangle = 0$. The last term in (8.62) describes the backreaction of the short-wavelength modes of ϕ onto the homogeneous mode and turns the deterministic slow roll evolution of ϕ into a stochastic process. The stochastic nature of Eq. (8.62) implies that the system can equivalently be described in terms of the Fokker-Planck equation for the probability density function $P(\phi, \phi_{\text{in}}, N)$

$$\frac{\partial P}{\partial N} = \frac{1}{3H_{\text{inf}}^2}\frac{\partial}{\partial \phi}(V_\phi P) + \frac{H_{\text{inf}}^2}{8\pi^2} P_{\phi\phi}, \tag{8.63}$$

where we take ϕ_{in} to be fixed at the onset of inflation. Once the solution to Eq. (8.63) is known, all relevant moments of the distribution can be computed:

$$\langle \phi^n \rangle(N) = \int d\phi\, \phi^n P(\phi, \phi_{\text{in}}, N)\,. \tag{8.64}$$

In what follows we will be interested in the first two moments: the mean, $\langle\phi\rangle$, and the mean square, $\langle\phi^2\rangle$, which allow us to determine the variance of the distribution $\sqrt{\langle\phi\rangle^2 - \langle\phi^2\rangle}$.

Given that the energy scale of quintessence is hierarchically smaller than the scale of inflation, the quintessence field is classically frozen during the inflationary epoch. It is only expected to thaw once the background Hubble parameter drops to around $H \sim m_\phi$, which should happen during the matter phase, after the Big-Bang. This should hold true regardless of the shape of the quintessential potential.

The existence of a vast hierarchy between the value of the Hubble parameter today and during inflation, $H_0^2 \simeq V_0/M_p^2 \ll H_{\text{inf}}^2$, implies that the quintessence field is a spectator during inflation, and that it is undergoing pure Brownian motion with Eq. (8.62) well approximated by

$$\phi' = \frac{H_{\text{inf}}}{2\pi}\xi\,; \tag{8.65}$$

or equivalently Eq. (8.63), by the one-dimensional heat equation

$$\frac{\partial P}{\partial N} = \frac{H_{\text{inf}}^2}{8\pi^2} P_{\phi\phi}. \tag{8.66}$$

This dominance of the stochastic effects over the classical evolution has severe consequences for the retention of memory of the initial conditions for the quintessence field. Exact solutions to Eq. (8.66) take the form (see e.g. [112])

$$P = \sqrt{\frac{2\pi}{N H_{\text{inf}}^2}} \exp\left(-\frac{2\pi^2}{N}\frac{(\phi - \phi_{\text{in}})^2}{H_{\text{inf}}^2}\right) \tag{8.67}$$

from which one can show that

$$\langle \phi \rangle = \phi_{\text{in}}, \tag{8.68}$$

i.e. the ensemble average is frozen at the specified initial value for the classical field ϕ_{in}, in accordance with the fact that classically the field is frozen by Hubble friction. One can also show that

$$\langle \phi^2 \rangle = \left(\frac{H_{\text{inf}}}{2\pi}\right)^2 N + \phi_{\text{in}}^2, \tag{8.69}$$

which implies

$$\sqrt{\langle \phi^2 \rangle - \langle \phi \rangle^2} = \frac{H_{\text{inf}}}{2\pi}\sqrt{N}\,. \tag{8.70}$$

Therefore in one efolding of inflationary expansion, the spectator field will be kicked on average by $H_{\text{inf}}/(2\pi)$. Depending on the sensitivity of a given hilltop to the choice of initial conditions, and on the exact value of H_{inf}, these stochastic effects can push the field away from the top of the hill, and into a region where it cannot account for the observed present day accelerated expansion.

From Eq. (8.67) we can compute the probability that a given choice of initial conditions survives the stochastic diffusion during a period of inflation. Setting $\phi_{\text{in}} = 0$ (assuming that this corresponds to the location of the maximum) and asking that after N efoldings of inflation ϕ remains within a distance Δ, we find that the survival probability is given by

$$\mathbb{P}(|\phi| \leq \Delta) = \int_{-\Delta}^{\Delta} d\phi\, P = \text{erf}\left(\sqrt{\frac{2\pi^2}{N}}\frac{\Delta}{H_{\text{inf}}}\right), \tag{8.71}$$

where erf is the complementary error function. For $N \gg 2\left(\frac{\pi\Delta}{H_{\text{inf}}}\right)^2$ we can approximate Eq. (8.71) by

$$\mathbb{P}(|\phi| \leq \Delta) \simeq 2\sqrt{\frac{2\pi}{N}}\frac{\Delta}{H_{\text{inf}}}. \tag{8.72}$$

Once again we see that if $H_{\text{inf}} \leq \Delta$ the memory of the choice of initial conditions is preserved for a long period. In Fig. 8.5 we plot the survival probability for various choices of Δ/H_{inf}.

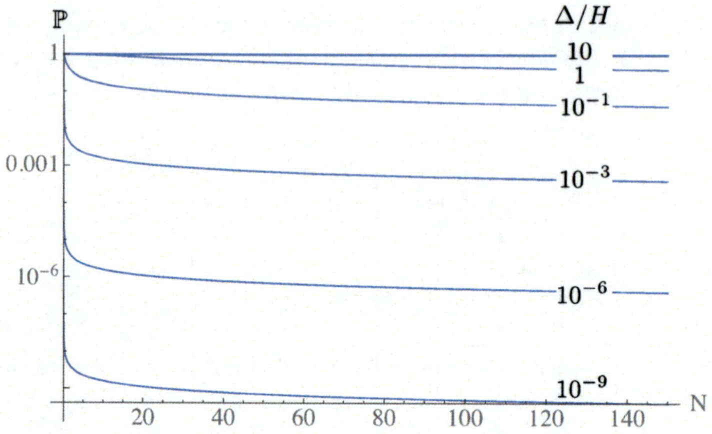

Fig. 8.5 Survival probability to remain within a distance Δ from the initial condition after N efoldings of inflation, as a function of the ratio $\Delta/H_{\rm inf}$

Axion hilltop quintessence

Stochastic effects on axionic hilltop quintessence models were studied in [74] and were pivotal in making the case for super-Planckian decay constants. Let us revisit these models in the light of the formalism reviewed above.

Recall that current bounds on primordial tensor modes imply that, during inflation, $H_{\rm inf} \lesssim 10^{-5} M_p$ and that stochastic effects become relevant when the width of the region around the maximum of V that gives rise to late time expansion is at, or below, the inflationary energy: $\Delta_{\rm max} \lesssim H_{\rm inf}$. In Fig. 8.6 we zoom in on the low f_a region of Fig. 8.3 and superimpose the constraints from $H_{\rm inf}$. We are led to the conclusion that axionic quintessence hilltop models with $f_a < 0.1\, M_p$ are subject to stochastic fluctuations that (depending on the inflationary energy scale) may push $\phi > \Delta_{\rm max}$ ruining the late time dynamics of those models. These estimates are in agreement with those of [74] and provide a worst case scenario. We note that a sharper statement can only be made with the knowledge of the exact energy scale of inflation.

In Fig. 8.7 we plot the solutions to the Langevin and Fokker-Planck equations for an axionic spectator with $H_{\rm inf} \simeq \Delta_{\rm max}$, where it is evident that stochastic effects lead to a loss of memory of initial conditions after $N > \mathcal{O}(10)$ efoldings of inflation.

The situation is even worse for $H_{\rm inf} > \Delta_{\rm max}$ since the probability of lying within a distance $\Delta_{\rm max}$ away from the maximum after $N \simeq 50-60$ efoldings of inflation turns out to be extremely small, as can be seen from (8.72) with $\Delta = \Delta_{\rm max}(f_a)$:

$$\mathbb{P}(|\phi| \leq \Delta_{\rm max}(f_a)) \simeq 2\sqrt{\frac{2\pi}{N}}\frac{\Delta_{\rm max}(f_a)}{H_{\rm inf}}, \tag{8.73}$$

where $\Delta_{\rm max}$ depends on the axion decay constant f_a (for $f_a \in [0.02, 0.1]\, M_p$, $\Delta_{\rm max}$ is very well approximated by (8.58)). For example, in the explicit LVS axion model of Sect. 8.3.1, Eqs. (8.53) and (8.58) give a decay constant $f_a \simeq 0.02\, M_p$ and a

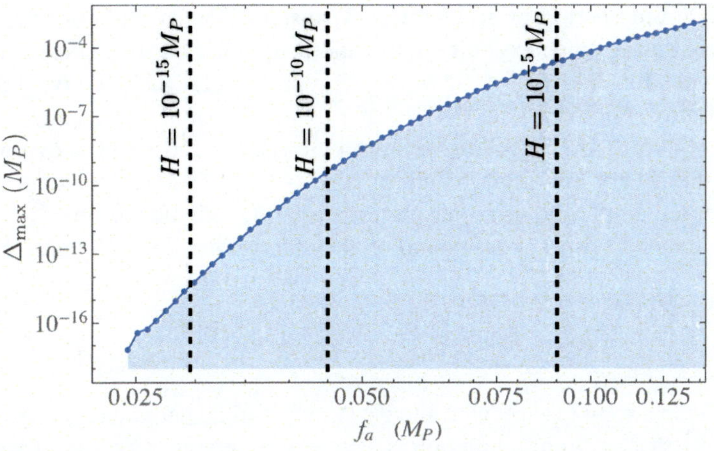

Fig. 8.6 Effect of quantum diffusion on the choice of initial conditions for axion hilltop quintessence models. Stochastic effects do not push the axion away from the hilltop region which yields late time acceleration only if $H_{\rm inf} \lesssim \Delta_{\max}(f_a)$

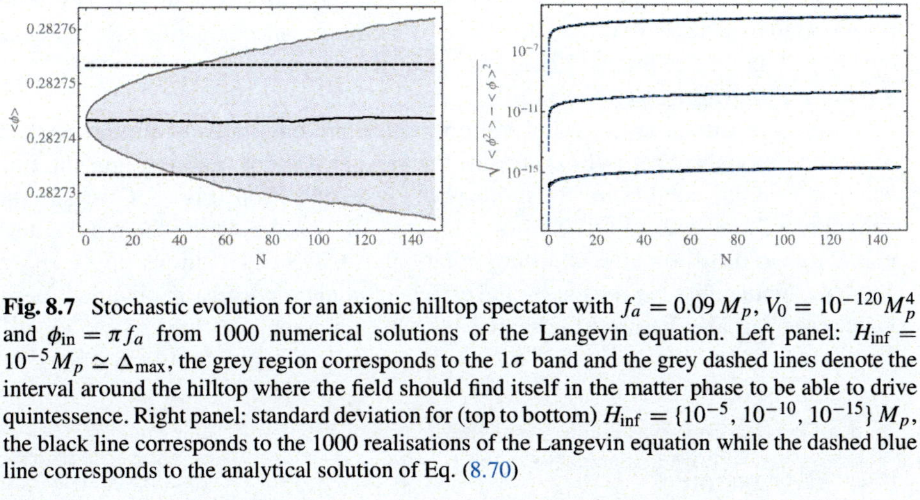

Fig. 8.7 Stochastic evolution for an axionic hilltop spectator with $f_a = 0.09\, M_p$, $V_0 = 10^{-120} M_p^4$ and $\phi_{\rm in} = \pi f_a$ from 1000 numerical solutions of the Langevin equation. Left panel: $H_{\rm inf} = 10^{-5} M_p \simeq \Delta_{\max}$, the grey region corresponds to the 1σ band and the grey dashed lines denote the interval around the hilltop where the field should find itself in the matter phase to be able to drive quintessence. Right panel: standard deviation for (top to bottom) $H_{\rm inf} = \{10^{-5}, 10^{-10}, 10^{-15}\}\, M_p$, the black line corresponds to the 1000 realisations of the Langevin equation while the dashed blue line corresponds to the analytical solution of Eq. (8.70)

maximum displacement, $\Delta_{\max} \simeq 2.4 \times 10^{-20}\, M_p$. Choosing the largest value of $H_{\rm inf}$ compatible with the lack of observation of primordial gravity waves, $H_{\rm inf} \simeq 2 \times 10^{-5}\, M_p$, and $N \simeq 50$ (8.73) would give $\mathbb{P}(|\phi| \leq \Delta_{\max}) \simeq 10^{-15}$. Clearly smaller values of $H_{\rm inf}$ would give a smaller survival probability. Notice from Fig. 8.3 that $\Delta_{\max}$ drops very quickly for smaller values of f_a, reducing the survival probability even further.

We therefore conclude that a viable axion hilltop quintessence model requires two crucial conditions: (1) a tuning of initial conditions close to the maximum which becomes more severe for smaller values of f_a; and (2) $H_{\rm inf} \lesssim \Delta_{\max}$, else stochastic effects will very quickly push the field away from the hilltop region compatible with a late time period of accelerated expansion. This second constraint turns out to be very strong since explicit computations of axion decay constants from string theory typically yield $f_a \lesssim 0.02\, M_p$ in the regime where the effective field theory is under control. For these small values of f_a, the maximum displacement is bounded as $\Delta_{\max} \lesssim 10^{-20}\, M_p$.[6] Inserting this into the second constraint above, we see that we require a very low scale of inflation, $H_{\inf} \lesssim 10^{-20}\, M_p$. When combined with the observed amplitude of scalar perturbations A_s, this bound on $H_{\inf}$ can then be translated into a severe upper bound on the inflationary slow roll parameter

$$\epsilon_V = \frac{1}{8\pi^2 A_s}\left(\frac{H_{\inf}}{M_p}\right)^2 \lesssim 10^{-35}. \tag{8.74}$$

This is in strong tension with the dS swampland conjecture (8.2) which requires $\mathcal{O}(1)$ values of ϵ_V (unless inflation is also realised extremely close to a maximum). Thus we conclude that axion hilltop quintessence would either be in the swampland or, if we ignore the refined dS swampland conjecture, would require a high tuning of initial conditions combined with a Hubble scale during inflation below $\mathcal{O}(1-10)$ MeV. Because these models are so contrived, we expect that dynamical dark energy, if supported by data, will have to be driven by a different mechanism, probably along the lines of axion alignment [72].

Saxion hilltop quintessence

The effects of diffusion in saxion hilltop models are qualitatively similar to those of axionic models. They will constrain the steeper hilltops, as these are the ones where the initial conditions are more severely tuned. From Fig. 8.4 we see that quadratic hilltops with $\eta_0 \gtrsim 70$ require $\Delta_{\max} \lesssim 10^{-5} M_p$ rendering them potentially vulnerable to diffusion effects during inflation, as CMB observations imply $H_{\inf} \lesssim 10^{-5} M_p$. Notice that the results for the effects of quantum diffusion obtained for the axion case can also be used for saxion hilltop quintessence as long as we identify $|\eta_0| = (M_p/f_a)^2$.

8.4 Discussion

With compelling observational evidence for dark energy [2, 113, 114], we cannot avoid the question of its microscopic origin. But should we be looking for a cosmological constant or quintessence? If the latter, then is it driven by a scalar or pseudo-scalar, on a shallow potential or at a hilltop? At present all options are obser-

[6] Numerically, we did not consider decay constants $f_a < 0.02\, M_p$ since the high degree of tuning of the initial conditions rapidly brings about numerical precision issues. This prevent us from explicitly determining $\Delta_{\max}$ for such low values of f_a, although the result can be obtained by extrapolating the validity of the formula Eq. (8.66). In any event, the precise estimate is not really needed for drawing our general conclusions.

vationally viable, but we can also ask which is easiest to build into a fundamental theory.

In this chapter, we have outlined several challenges facing string theory models of quintessence focusing on effective field theories where perturbation theory is under numerical control—i.e. where the dilaton, $\mathrm{Re}(S) \gg 1$, and the volume mode, $\mathcal{V} \gg 1$, are large enough to trust both the string loop and the α' expansion. This is arguably the most interesting region of moduli space since deep in the bulk, where $\mathrm{Re}(S) \sim \mathcal{V} \sim \mathcal{O}(1)$, one would need a full knowledge of the whole quantum theory, while at boundary of the moduli space, where $\mathrm{Re}(S) \to \infty$ and $\mathcal{V} \to \infty$, there is strong evidence indicating the absence of both dS vacua [8] and a viable quintessence dynamics [46–49].

Some of the problems of dynamical dark energy models are shared with the pure cosmological constant (like the smallness of H_0), while others are particular to quintessence (like constraints from fifth forces, tuning of initial conditions and radiative stability of the mass of the quintessence field). Here we were particularly concerned with the destabilisation of the volume modulus during inflation (the KL problem applied to quintessence) and generating the large hierarchy between the scale of the universe today and during inflation. We have argued that the need to overcome all of these challenges has singled out a preferred model building scenario for dynamical dark energy. The leading order contributions to the scalar potential should yield a vacuum with the following properties:

1. it should admit a flat direction in order to decouple the dark energy scale from the inflationary scale;
2. this flat direction should be axionic. This is because saxions are already lifted at perturbative level without being able to generate the required hierarchy between H_0 and H_{inf}. Axions, in contrast, develop a potential via highly-suppressed non-perturbative effects;
3. it should be (nearly) Minkowski since otherwise subleading corrections would not be able to push it up to the positive energies required to drive an epoch of accelerated expansion;
4. it should break supersymmetry in order to decouple the gravitino mass from the dark energy scale.

It is interesting to combine these results with the swampland dS conjecture that would rule out dS vacua. From a model-building perspective, dS vacua look qualitatively the same as non-supersymmetric Minkowski, leaving dynamical dark energy as the only explanation for the present acceleration of the universe. However we have found that it is extremely hard to realise a working model of quintessence in any scenario which would be in agreement with the swampland dS conjecture, like moderately sloped runaways, or supersymmetric AdS or Minkowski vacua. This tension raises some doubts on the validity of the swampland dS conjecture since it would imply that quantum gravity is in contradiction with observations.

At this point it is tempting to favour the humble cosmological constant as the simplest empirical model of dark energy: it fits the available data and avoids the additional complications associated with quintessence. However, it is important to note that quintessence can open up opportunities to solve other cosmological problems. For example, in Chap. 5, it was shown how dynamical models of quintessence in string theory may shed new light on the cosmological coincidence problem [43, 44]. An evolving scalar on cosmological scales may also allow for self adjustment mechanisms to address the naturalness problems associated with vacuum energy (see [115, 116] for relevant no go theorems, and [117] for a recent way around them). But perhaps most importantly, future observations may rule out the cosmological constant as the driver of late time acceleration.

If this were indeed the case, our analysis provides guidance for successful quintessence model building in string theory. In fact, we studied axion hilltop quintessence in detail since vanilla string compactifications lead to axion decay constants at least two orders of magnitude below the Planck scale. We found that hilltop models are rather contrived since, even if the initial conditions are tuned very close to the maximum, quantum diffusion effects during inflation would kick the quintessence field away from the accelerating region close to the maximum, unless the Hubble scale during inflation is extremely low, $H_{\rm inf} \lesssim \mathcal{O}(1-10)$ MeV. Of course, one could envisage a scenario where a suitable coupling between the inflaton and the quintessence field makes the latter heavy during inflation, thereby suppressing any stochastic effect. However, after the end of inflation, the inflaton would typically settle down at the minimum of its potential, reaching its present day value. Quantum diffusion would then still play an important rôle in the reheating phase and after, implying that the results of Sect. 8.3.3 would still hold qualitatively after inflation with $H_{\rm inf}$ replaced by the Hubble scale during a given epoch. We conclude that a more promising avenue to build a working model of dynamical dark energy is to rely on alignment mechanisms to obtain an effective axion decay constant which is trans-Planckian [72].

As stated earlier, for dynamical dark energy, we first seek a scenario where the volume is stabilised at leading order to a vacuum that is uplifted to Minkowski. Non-perturbative corrections can then be exploited to drive dark energy at the correct scale. Although it might seem a little uneconomical to uplift and then do quintessence, if dark energy turns out to be dynamical both steps may be necessary to explain the present state of the universe in the context of string compactifications.

In truth, both the cosmological constant and quintessence face formidable challenges from the perspective of consistent model building in string theory, while remaining perfectly compatible with observational constraints. It behooves us to better understand the limitations imposed by perturbative string theory in both cases. Indeed, does a microscopic understanding of dark energy require input from non-perturbative strings, through string field theory, or M-theory? Since this may be a question of properly understanding the vacuum structure of the theory, this seems like a reasonable possibility.

References

1. Obied G, Ooguri H, Spodyneiko L, Vafa C. De Sitter space and the swampland. https://arxiv.org/abs/1806.08362
2. Planck Collaboration (2020) Planck 2018 results. VI. Cosmological parameters. Astron Astrophys 641:A6. https://doi.org/10.1051/0004-6361/201833910, http://arxiv.org/abs/1807.06209
3. Planck Collaboration (2020) Planck 2018 results. X. Constraints on inflation. Astron Astrophys 641:A10. https://doi.org/10.1051/0004-6361/201833887, http://arxiv.org/abs/1807.06211
4. Amendola L et al (2018) Cosmology and fundamental physics with the Euclid satellite. Living Rev Rel 21:2. https://doi.org/10.1007/s41114-017-0010-3, http://arxiv.org/abs/1606.00180
5. Denef F, Hebecker A, Wrase T (2018) de Sitter swampland conjecture and the Higgs potential. Phys Rev D 98:086004. https://doi.org/10.1103/PhysRevD.98.086004, http://arxiv.org/abs/1807.06581
6. Conlon JP (2018) The de Sitter swampland conjecture and supersymmetric AdS vacua. Int J Mod Phys A 33:1850178. https://doi.org/10.1142/S0217751X18501786, http://arxiv.org/abs/1808.05040
7. Garg SK, Krishnan C (2019) Bounds on slow roll and the de Sitter swampland. JHEP 11:075. https://doi.org/10.1007/JHEP11(2019)075, http://arxiv.org/abs/1807.05193
8. Ooguri H, Palti F, Shiu G, Vafa C (2019) Distance and de Sitter conjectures on the swampland. Phys Lett B 788:180. https://doi.org/10.1016/j.physletb.2018.11.018, http://arxiv.org/abs/1810.05506
9. Louis J, Rummel M, Valandro R, Westphal A (2012) Building an explicit de Sitter. JHEP 10:163. https://doi.org/10.1007/JHEP10(2012)163, http://arxiv.org/abs/1208.3208
10. Cicoli M, Krippendorf S, Mayrhofer C, Quevedo F, Valandro R (2012) D-branes at del Pezzo singularities: global embedding and moduli stabilisation. JHEP 09:019. https://doi.org/10.1007/JHEP09(2012)019, http://arxiv.org/abs/1206.5237
11. Cicoli M,Maharana A, Quevedo F, Burgess CP (2012) De Sitter string Vacua from dilaton-dependent non-perturbative effects. JHEP 06:011. https://doi.org/10.1007/JHEP06(2012)011, http://arxiv.org/abs/1203.1750
12. Cicoli M, Klevers D, Krippendorf S, Mayrhofer C, Quevedo F, Valandro R (2014) Explicit de Sitter flux vacua for global string models with chiral matter. JHEP05:001. https://doi.org/10.1007/JHEP05(2014)001, http://arxiv.org/abs/1312.0014
13. Blåbäck J, Danielsson U, Dibitetto G (2013) Fully stable dS vacua from generalised fluxes. JHEP 08:054. https://doi.org/10.1007/JHEP08(2013)054, http://arxiv.org/abs/1301.7073
14. Braun AP, Rummel M, Sumitomo Y, Valandro R (2015) De Sitter vacua from a D-term generated racetrack potential in hypersurface Calabi-Yau compactifications. JHEP 12:033. https://doi.org/10.1007/JHEP12(2015)033, http://arxiv.org/abs/1509.06918
15. Cribiori N, Kallosh R, Linde A, Roupec C (2020) de Sitter minima from M theory and string theory. Phys Rev D 101:046018. https://doi.org/10.1103/PhysRevD.101.046018, http://arxiv.org/abs/1912.02791
16. Antoniadis I, Chen Y, Leontaris GK (2020) Logarithmic loop corrections, moduli stabilisation and de Sitter vacua in string theory. JHEP 01:149. https://doi.org/10.1007/JHEP01(2020)149, http://arxiv.org/abs/1909.10525
17. Crinò C, Quevedo F, Valandro R (2021) On de Sitter string vacua from anti-D3-branes in the large volume scenario. JHEP 03:258. https://doi.org/10.1007/JHEP03(2021)258, http://arxiv.org/abs/2010.15903
18. Cicoli M, Etxebarria ING, Quevedo F, Schachner A, Shukla P, Valandro R (2021) The standard model quiver in de Sitter string compactifications. JHEP 08:109. https://doi.org/10.1007/JHEP08(2021)109, http://arxiv.org/abs/2106.11964
19. Kachru S, Kallosh R, Linde AD, Trivedi SP (2003) De Sitter vacua in string theory. Phys Rev D 68:046005. https://doi.org/10.1103/PhysRevD.68.046005, http://arxiv.org/abs/hep-th/0301240

20. Balasubramanian V, Berglund P, Conlon JP, Quevedo F (2005) Systematics of moduli stabilisation in Calabi-Yau flux compactifications. JHEP 03:007. https://doi.org/10.1088/1126-6708/2005/03/007, http://arxiv.org/abs/hep-th/0502058arXIV:hep-th/0502058
21. Conlon JP, Quevedo F, Suruliz K (2005) Large-volume flux compactifications: moduli spectrum and D3/D7 soft supersymmetry breaking. JHEP 08:007. https://doi.org/10.1088/1126-6708/2005/08/007, http://arxiv.org/abs/hep-th/0505076
22. Cicoli M, Conlon JP, Quevedo F (2008)General analysis of LARGE volume scenarios with string loop moduli stabilisation. JHEP 10:105. https://doi.org/10.1088/1126-6708/2008/10/105, http://arxiv.org/abs/0805.1029
23. Becker K, Becker M, Haack M, Louis J (2002) Supersymmetry breaking and alpha-prime corrections to flux induced potentials. JHEP 06:060. https://doi.org/10.1088/1126-6708/2002/06/060, http://arxiv.org/abs/hep-th/0204254
24. Berg M, Haack M, Kors B (2005) String loop corrections to Kahler potentials in orientifolds. JHEP 11:030. https://doi.org/10.1088/1126-6708/2005/11/030, http://arxiv.org/abs/hep-th/0508043
25. von Gersdorff G, Hebecker A (2005) Radius stabilization by two-loop Casimir energy. Nucl Phys B 720:211. https://doi.org/10.1016/j.nuclphysb.2005.06.001,http://arxiv.org/abs/hep-th/0504002
26. Berg M, Haack M, Pajer E (2007) Jumping through loops: on soft terms from large volume compactifications. JHEP 09:031. https://doi.org/10.1088/1126-6708/2007/09/031, http://arxiv.org/abs/0704.0737
27. Cicoli M, Conlon JP, Quevedo F (2008) Systematics of string loop corrections in Type IIB Calabi-Yau flux compactifications. JHEP 01:052. https://doi.org/10.1088/1126-6708/2008/01/052, http://arxiv.org/abs/0708.1873
28. Ciupke D, Louis J, Westphal A (2015) Higher-derivative supergravity and moduli stabilization. JHEP 10:094. https://doi.org/10.1007/JHEP10(2015)094, http://arxiv.org/abs/1505.03092
29. Antoniadis I, Chen Y, Leontaris GK (2018) Perturbative moduli stabilisation in type IIB/F-theory framework. Eur Phys J C 78:766. https://doi.org/10.1140/epjc/s10052-018-6248-4, http://arxiv.org/abs/1803.08941
30. Blumenhagen R, Cvetic M, Kachru S, Weigand T (2009) D-brane instantons in Type II orientifolds. Ann Rev Nucl Part Sci 59:269. https://doi.org/10.1146/annurev.nucl.010909.083113, http://arxiv.org/abs/0902.3251
31. Burgess CP, Cicoli M, Ciupke D, Krippendorf S, Quevedo F (2020) UV shadows in EFTs: accidental symmetries, robustness and no-scale supergravity. Fortsch Phys 68:2000076. https://doi.org/10.1002/prop.202000076, http://arxiv.org/abs/2006.06694
32. Cicoli M, Quevedo F, Savelli R, Schachner A, Valandro R (2021) Systematics of the α' expansion in F-theory. JHEP 08:099. https://doi.org/10.1007/JHEP08(2021)099, http://arxiv.org/abs/2106.04592
33. Maldacena JM, Nunez C (2001) Supergravity description of field theories on curved manifolds and a no go theorem. Int J Mod Phys A 16:822. https://doi.org/10.1142/S0217751X01003937, http://arxiv.org/abs/hep-th/0007018
34. Green SR, Martinec EJ, Quigley C, Sethi S (2012) Constraints on string cosmology. Class Quant Grav 29:075006. https://doi.org/10.1088/0264-9381/29/7/075006, http://arxiv.org/abs/1110.0545
35. Gautason FF, Junghans D, Zagermann M (2012) On cosmological constants from α'-corrections. JHEP 06:029. https://doi.org/10.1007/JHEP06(2012)029, http://arxiv.org/abs/1204.0807
36. Dasgupta K, Gwyn R, McDonough E, Mia M, Tatar R (2014) de Sitter vacua in Type IIB string theory: classical solutions and quantum corrections. JHEP 07:054. https://doi.org/10.1007/JHEP07(2014)054, http://arxiv.org/abs/1402.5112
37. Kutasov D, Maxfield T, Melnikov I, Sethi S (2015) Constraining de Sitter space in string theory. Phys Rev Lett 115:071305. https://doi.org/10.1103/PhysRevLett.115.071305, http://arxiv.org/abs/1504.00056

38. Quigley C (2015) Gaugino condensation and the cosmological constant. JHEP **06**:104. https://doi.org/10.1007/JHEP06(2015)104, http://arxiv.org/abs/1504.00652
39. Dine M, Law-Smith JAP, Sun S, Wood D, Yu Y (2021) Obstacles to constructing de Sitter space in string theory. JHEP 02:050. https://doi.org/10.1007/JHEP02(2021)050, http://arxiv.org/abs/2008.12399
40. Montero M, Van Riet T, Venken G (2020) A dS obstruction and its phenomenological consequences. JHEP 05:114. https://doi.org/10.1007/JHEP05(2020)114, http://arxiv.org/abs/2001.11023
41. Cunillera F, Emond WT, Lehébel A, Padilla A (2022) Quadratic curvature corrections to stringy effective actions and the absence of de Sitter vacua. JHEP 02:012. https://doi.org/10.1007/JHEP02(2022)012,http://arxiv.org/abs/2112.05771
42. Bedroya A, Vafa C (2020) Trans-Planckian censorship and the swampland. JHEP 09:123. https://doi.org/10.1007/JHEP09(2020)123, http://arxiv.org/abs/1909.11063
43. Zlatev I, Wang LM, Steinhardt PJ (1999) Quintessence, cosmic coincidence, and the cosmological constant. Phys Rev Lett 82:896. https://doi.org/10.1103/PhysRevLett.82.896, http://arxiv.org/abs/astro-ph/9807002
44. Velten HES, vom Marttens RF, Zimdahl W (2014) Aspects of the cosmological "coincidence problem". Eur Phys J C. 74:3160. https://doi.org/10.1140/epjc/s10052-014-3160-4, http://arxiv.org/abs/1410.2509
45. Cunillera F, Padilla A (2021) A stringy perspective on the coincidence problem. JHEP 10:055. https://doi.org/10.1007/JHEP10(2021)055, http://arxiv.org/abs/2105.03426
46. Cicoli M, Cunillera F, Padilla A, Pedro FG (2022) Quintessence and the swampland: the parametrically controlled regime of moduli space. Fortsch Phys 70. https://doi.org/10.1002/prop.202200009, http://arxiv.org/abs/2112.10779
47. Hertzberg MP, Kachru S, Taylor W, Tegmark M (2007) Inflationary constraints on Type IIA string theory. JHEP 12:095. https://doi.org/10.1088/1126-6708/2007/12/095, http://arxiv.org/abs/0711.2512
48. Garg SK, Krishnan C, Zaid Zaz M (2019) Bounds on slow roll at the boundary of the landscape. JHEP 03:029. https://doi.org/10.1007/JHEP03(2019)029, http://arxiv.org/abs/1810.09406
49. Valeixo Bento B, Chakraborty D, Parameswaran SL, Zavala I (2020) Dark energy in string theory. PoS CORFU2019:123. https://doi.org/10.22323/1.376.0123, http://arxiv.org/abs/2005.10168
50. Cicoli M, Dibitetto G, Pedro FG ()2020 New accelerating solutions in late-time cosmology. Phys Rev D 101:103524. https://doi.org/10.1103/PhysRevD.101.103524, http://arxiv.org/abs/2002.02695
51. Cicoli M, Dibitetto G, Pedro FG (2020) Out of the swampland with multifield quintessence?. JHEP 10:035. https://doi.org/10.1007/JHEP10(2020)035, http://arxiv.org/abs/2007.11011
52. Brinkmann M, Cicoli M, Dibitetto G, Pedro FG. Stringy multifield quintessence and the Swampland. http://arxiv.org/abs/2206.10649
53. Cicoli M, De Alwis S, Maharana A, Muia F, Quevedo F (2019) De Sitter versus quintessence in string theory. Fortsch Phys 67:1800079. https://doi.org/10.1002/prop.201800079, http://arxiv.org/abs/1808.08967
54. Hebecker A, Skrzypek T, Wittner M (2019) The F-term problem and other challenges of stringy quintessence. JHEP 11:134. https://doi.org/10.1007/JHEP11(2019)134, http://arxiv.org/abs/1909.08625
55. Will CM (2014) The confrontation between general relativity and experiment. Living Rev Rel 17:4. https://doi.org/10.12942/lrr-2014-4, http://arxiv.org/abs/1403.7377
56. Colgáin EO, Yavartanoo H (2019) Testing the swampland: H_0 tension. Phys Lett B 797:134907.https://doi.org/10.1016/j.physletb.2019.134907, http://arxiv.org/abs/1905.02555
57. Banerjee A, Cai H, Heisenberg L, Colgáin EO, Sheikh-Jabbari MM, Yang T (2021) Hubble sinks in the low-redshift swampland. Phys Rev D 103:L081305. https://doi.org/10.1103/PhysRevD.103.L081305, http://arxiv.org/abs/2006.00244

58. Riess AG et al. A comprehensive measurement of the local value of the Hubble constant with 1 km/s/Mpc uncertainty from the Hubble space telescope and the SH0ES team. http://arxiv.org/abs/2112.04510
59. Kallosh R, Linde AD (2004) Landscape, the scale of SUSY breaking, and inflation. JHEP 12:004. https://doi.org/10.1088/1126-6708/2004/12/004, http://arxiv.org/abs/hep-th/0411011
60. Cicoli M, Muia F, Pedro FG (2015) Microscopic origin of volume modulus inflation. JCAP 12:040. https://doi.org/10.1088/1475-7516/2015/12/040, http://arxiv.org/abs/1509.07748
61. Cicoli M, Pedro FG, Tasinato G (2012) Natural quintessence in string theory. JCAP 07:044. https://doi.org/10.1088/1475-7516/2012/07/044, http://arxiv.org/abs/1203.6655
62. Akrami Y, Kallosh R, Linde A, Vardanyan V (2019) The landscape, the swampland and the era of precision cosmology. Fortsch Phys 67:1800075. https://doi.org/10.1002/prop.201800075, http://arxiv.org/abs/1808.09440arXIV:1808.09440
63. Agrawal P, Obied G, Steinhardt PJ, Vafa C (2018) On the cosmological implications of the string swampland. Phys Lett B 784:271. https://doi.org/10.1016/j.physletb.2018.07.040, http://arxiv.org/abs/1806.09718
64. Ferrer F, Grifols JA (1997) Constraints on the mass of the superlight gravitino from the muon anomaly. Phys Rev D 56:7466. https://doi.org/10.1103/PhysRevD.56.7466, http://arxiv.org/abs/hep-ph/9704444arXIV:hep-ph/9704444
65. Brignole A, Feruglio F, Zwirner F (1998)Signals of a superlight gravitino at e^+e^- colliders when the other superparticles are heavy. Nucl Phys B 516:13. https://doi.org/10.1016/S0550-3213(97)00825-0, http://arxiv.org/abs/hep-ph/9711516
66. Brignole A, Feruglio F, Mangano ML, Zwirner F (1998) Signals of a superlight gravitino at hadron colliders when the other superparticles are heavy. Nucl Phys B 526:136. https://doi.org/10.1016/S0550-3213(98)00254-5, http://arxiv.org/abs/hep-ph/9801329
67. Kawasaki M, Kohri K, Moroi T, Yotsuyanagi A (2008) Big-bang nucleosynthesis and gravitino, Phys Rev D 78:065011. https://doi.org/10.1103/PhysRevD.78.065011, http://arxiv.org/abs/0804.3745
68. Burgess CP, Cicoli M, Quevedo F, Williams M (2014) Inflating with large effective fields. JCAP 11: 045. https://doi.org/10.1088/1475-7516/2014/11/045, http://arxiv.org/abs/1404.6236
69. Burgess CP, Cicoli M, de Alwis S, Quevedo F (2016) Robust inflation from fibrous strings. JCAP 05:032. https://doi.org/10.1088/1475-7516/2016/05/032, http://arxiv.org/abs/1603.06789
70. Cicoli M, Goodsell M, Ringwald A (2012) The Type IIB string axiverse and its low-energy phenomenology. JHEP 10:146. https://doi.org/10.1007/JHEP10(2012)146, http://arxiv.org/abs/1206.0819
71. Arkani-Hamed N, Motl L, Nicolis A, Vafa C (2007) The string landscape, black holes and gravity as the weakest force. JHEP 06:060. https://doi.org/10.1088/1126-6708/2007/06/060, http://arxiv.org/abs/hep-th/0601001
72. Kim JE, Nilles HP, Peloso M (2005) Completing natural inflation. JCAP 01:005. https://doi.org/10.1088/1475-7516/2005/01/005, http://arxiv.org/abs/hep-ph/0409138
73. Dimopoulos S, Kachru S, McGreevy J, Wacker JG (2008) N-flation. JCAP 08:003. https://doi.org/10.1088/1475-7516/2008/08/003, http://arxiv.org/abs/hep-th/0507205
74. Kaloper N, Sorbo L (2006) Of pNGB quintessence. JCAP 04:007. https://doi.org/10.1088/1475-7516/2006/04/007, http://arxiv.org/abs/astro-ph/0511543
75. Blanco-Pillado JJ, Burgess CP, Cline JM, Escoda C, Gomez-Reino M, Kallosh R et al (2004) Racetrack inflation. JHEP 11:063. https://doi.org/10.1088/1126-6708/2004/11/063, http://arxiv.org/abs/hep-th/0406230
76. Choi K (2000) String or M theory axion as a quintessence. Phys Rev D 62:043509. https://doi.org/10.1103/PhysRevD.62.043509, http://arxiv.org/abs/hep-ph/9902292
77. Kaloper N, Sorbo L (2009) Where in the string landscape is quintessence. Phys Rev D 79:043528. https://doi.org/10.1103/PhysRevD.79.043528, http://arxiv.org/0810.5346

78. Panda S, Sumitomo Y, Trivedi SP (2011) Axions as quintessence in string theory. Phys Rev D 83:083506. https://doi.org/10.1103/PhysRevD.83.083506, http://arxiv.org/abs/1011.5877
79. Kamionkowski M, Pradler J, Walker DGE (2014) Dark energy from the string axiverse. Phys Rev Lett 113:251302. https://doi.org/10.1103/PhysRevLett.113.251302, http://arxiv.org/abs/1409.0549
80. Olguin-Trejo Y, Parameswaran SL, Tasinato G, Zavala I (2019) Runaway quintessence, out of the swampland. JCAP 01:031. https://doi.org/10.1088/1475-7516/2019/01/031, http://arxiv.org/abs/1810.08634
81. PARTICAL DATA GROUP Collaboration (2020) Review of particle physics. PTEP 2020:083C01. https://doi.org/10.1093/ptep/ptaa104
82. Kapner DJ, Cook TS, Adelberger EG, Gundlach JH, Heckel BR, Hoyle CD et al (2007) Tests of the gravitational inverse-square law below the dark-energy length scale. Phys Rev Lett 98:021101. https://doi.org/10.1103/PhysRevLett.98.021101, http://arxiv.org/abs/hep-ph/0611184
83. Damour T, Donoghue JF (2010) Equivalence principle violations and couplings of a light dilaton. Phys Rev D 82:084033. https://doi.org/10.1103/PhysRevD.82.084033, http://arxiv.org/abs/1007.2792
84. Acharya BS, Maharana A, Muia F (2019) Hidden sectors in string theory: kinetic mixings, fifth forces and quintessence. JHEP 03:048. https://doi.org/10.1007/JHEP03(2019)048, http://arxiv.org/abs/1811.10633
85. Cicoli M, Burgess CP, Quevedo F (2011) Anisotropic modulus stabilisation: strings at LHC scales with micron-sized extra dimensions. JHEP 10:119. https://doi.org/10.1007/JHEP10(2011)119, http://arxiv.org/abs/1105.2107
86. Burgess CP, Quevedo F. Axion homeopathy: screening dilaton interactions. http://arxiv.org/abs/2110.10352
87. Burgess CP, Dineen D, Quevedo F. Yoga dark energy: natural relaxation and other dark implications of a supersymmetric gravity sector. http://arxiv.org/abs/2111.07286
88. Buchmuller W, Hamaguchi K, Lebedev O, Ratz M (2005) Maximal temperature in flux compactifications. JCAP 01:004. https://doi.org/10.1088/1475-7516/2005/01/004, http://arxiv.org/abs/hep-th/0411109
89. Anguelova L, Calo V, Cicoli M (2009) LARGE volume string compactifications at finite temperature. JCAP 10:025. https://doi.org/10.1088/1475-7516/2009/10/025, http://arxiv.org/abs/0904.0051
90. Conlon JP, Kallosh R, Linde SD, Quevedo F (2008) Volume modulus inflation and the gravitino mass problem. JCAP 09:011. https://doi.org/10.1088/1475-7516/2008/09/011, http://arxiv.org/abs/0806.0809
91. Hsu JP, Kallosh R, Prokushkin S (2003) On brane inflation with volume stabilization. JCAP 12:009. https://doi.org/10.1088/1475-7516/2003/12/009, http://arxiv.org/abs/hep-th/0311077
92. Aparicio L, Cicoli M, Krippendorf S, Maharana A, Muia F, Quevedo F (2014) Sequestered de Sitter string scenarios: soft-terms. JHEP 11:071. https://doi.org/10.1007/JHEP11(2014)071, http://arxiv.org/abs/1409.1931
93. Pitrou C, Coc A, Uzan J-P, Vangioni E (2018) Precision big bang nucleosynthesis with improved Helium-4 predictions. Phys Rept 754:1. https://doi.org/10.1016/j.physrep.2018.04.005, http://arxiv.org/abs/1801.08023
94. Davidson S, Losada M, Riotto A (2000) A new perspective on baryogenesis. Phys Rev Lett 84:4284. https://doi.org/10.1103/PhysRevLett.84.4284, http://arxiv.org/abs/hep-ph/0001301
95. He T, Kachru S, Westphal A (2010) Gravity waves and the LHC: towards high-scale inflation with low-energy SUSY. JHEP 06:065. https://doi.org/10.1007/JHEP06(2010)065, https://arxiv.org/abs/1003.4265
96. Hellerman S, Kaloper N, Susskind L (2001) String theory and quintessence. JHEP 06:003. https://doi.org/10.1088/1126-6708/2001/06/003, http://arxiv.org/abs/hep-th/0104180
97. Cicoli M, Kreuzer M, Mayrhofer C (2012) Toric K3-fibred Calabi-Yau manifolds with del Pezzo divisors for string compactifications. JHEP 02:002. https://doi.org/10.1007/JHEP02(2012)002, http://arxiv.org/abs/1107.0383

98. Cicoli M, Ciupke D, Mayrhofer C, Shukla P (2018) A geometrical upper bound on the inflaton range. JHEP 05:001. https://doi.org/10.1007/JHEP05(2018)001, http://arxiv.org/abs/1801.05434
99. Cicoli M, Quevedo F, Valandro R (2016) De Sitter from T-branes. JHEP03:141. https://doi.org/10.1007/JHEP03(2016)141, http://arxiv.org/abs/1512.04558
100. Cicoli M, Mayrhofer C, Valandro R (2012) Moduli stabilisation for chiral global models. JHEP 02:062. https://doi.org/10.1007/JHEP02(2012)062, http://arxiv.org/abs/1110.3333
101. Cicoli M, Krippendorf S, Mayrhofer C, Quevedo F, Valandro R (2013) D3/D7 branes at singularities: constraints from global embedding and moduli stabilisation. JHEP 07:150. https://doi.org/10.1007/JHEP07(2013)150, http://arxiv.org/abs/1304.0022
102. Cicoli M, Garcìa-Etxebarria IN, Mayrhofer C, Quevedo F, Shukla P, Valandro R, Global orientifolded quivers with inflation. JHEP 11:134. https://doi.org/10.1007/JHEP11(2017)134, http://arxiv.org/abs/1706.06128
103. Danielsson UH, Van Riet T (2018) What if string theory has no de Sitter vacua? Int J Mod Phys D 27:1830007. https://doi.org/10.1142/S0218271818300070, http://arxiv.org/abs/1804.01120
104. Cicoli M, Schachner A, Shukla P. Systematics of Type IIB moduli stabilisation with odd axions. http://arxiv.org/abs/2109.14624
105. Cicoli M, Guidetti V, Righi N, Westphal A. Fuzzy dark matter candidates from string theory. http://arxiv.org/abs/2110.02964
106. Shelton J, Taylor W, Wecht B (2005), Nongeometric flux compactifications. JHEP 10:085. https://doi.org/10.1088/1126-6708/2005/10/085, http://arxiv.org/abs/hep-th/0508133
107. Benmachiche I, Grimm TW (2006) Generalized N = 1 orientifold compactifications and the Hitchin functionals. Nucl Phys B 748:200. https://doi.org/10.1016/j.nuclphysb.2006.05.003, http://arxiv.org/abs/hep-th/0602241
108. Plauschinn E (2019) Non-geometric backgrounds in string theory. Phys Rept 798:1. https://doi.org/10.1016/j.physrep.2018.12.002, http://arxiv.org/abs/1811.11203
109. Scherrer RJ, Sen AA (2008) Thawing quintessence with a nearly flat potential. Phys Rev D 77:083515. https://doi.org/10.1103/PhysRevD.77.083515, http://arxiv.org/abs/0712.3450
110. Dutta S, Scherrer RJ (2008) Hilltop quintessence. Phys Rev D 78:123525. https://doi.org/10.1103/PhysRevD.78.123525, http://arxiv.org/abs/0809.4441
111. Hardwick RJ, Vennin V, Byrnes CT, Torrado J, Wands D (2017) The stochastic spectator. JCAP 10:018. https://doi.org/10.1088/1475-7516/2017/10/018, http://arxiv.org/abs/1701.06473
112. Uhlenbeck GE, Ornstein LS (1930) On the theory of the Brownian motion. Phys Rev 36:823. https://doi.org/10.1103/PhysRev.36.823
113. Supernove Cosmology Project Collaboration (1999) Measurements of Ω and Λ from 42 high redshift supernovae. Astrophys J 517:565. https://doi.org/10.1086/307221, http://arxiv.org/abs/astro-ph/9812133
114. Supernova Search Team Collaboration (1998) Observational evidence from supernovae for an accelerating universe and a cosmological constant. Astron J 116:1009. https://doi.org/10.1086/300499, http://arxiv.org/abs/astro-ph/9805201
115. Weinberg S (1987) Anthropic bound on the cosmological constant. Phys Rev Lett 59:2607. https://doi.org/10.1103/PhysRevLett.59.2607
116. Niedermann F, Padilla A (2017) Gravitational mechanisms to self-tune the cosmological constant: obstructions and ways forward. Phys Rev Lett 119:251306. https://doi.org/10.1103/PhysRevLett.119.251306, http://arxiv.org/abs/1706.04778
117. Copeland EJ, Ghataore S, Niedermann F, Padilla A. Generalised scalar-tensor theories and self-tuning. http://arxiv.org/abs/2111.11448

Chapter 9
Gauss-Bonnet Curvature Corrections and the Absence of de Sitter Vacua

In the previous chapter, we described a number of phenomenological impediments to quintessence model building. Many of the arguments we described rely on compactifications of string theory on Calabi-Yau manifolds, originally chosen to preserve some supersymmetry in the external space. Of course, to recover de Sitter space, supersymmetry must be broken completely,[1] so it is worth exploring string compactifications on manifolds that are not Calabi-Yau, at least in the rare cases where one can solve the equations and retain some calculational control. To this end, in [6], the authors took a generic class of string inspired models and studied compactifications on a p-sphere with internal p-form fluxes. They identified a parametric bound (henceforth known as the Montero-van Riet-Venken (MvRV) bound), that needed to hold for de Sitter solutions to exist. The MvRV bound corresponds to a remarkably simple inequality between the slope of the potential and the derivative of a p-form gauge coupling. A number of stringy models were studied where it could be explicitly shown that the inequality is not satisfied, ruling out de Sitter solutions in the most realistic set-ups.

In this section, we expand on the work of [6] by introducing higher order curvature corrections in the form of a Gauss-Bonnet contribution to the action, as well as generalising the set-up to contain internal or external fluxes with spherical or toroidal internal geometries. The goal is to establish whether the curvature corrections strengthen or weaken the MvRV bound of [6] for existence of de Sitter vacua. We will consider actions of the following form:

$$S = \int \mathrm{d}^D x \sqrt{-g} \left[\frac{R}{2\kappa_D^2} - \frac{1}{2}(\partial\varphi)^2 - V(\varphi) - g(\varphi)\,|H_k|^2 + f(\varphi)\mathcal{G} \right], \tag{9.1}$$

[1] We note that supersymmetry may be realised non-linearly on de Sitter space whenever the symmetry is broken spontaneously, as in [1–5].

F. Cunillera, *Dark Energy*, Springer Theses,
https://doi.org/10.1007/978-3-031-21468-4_9

where $\mathcal{G}$ is the Gauss-Bonnet term, cf. Eq. (9.11), and take our ansatz manifold to be the warped product $\mathcal{M} = \Sigma_n \rtimes_\rho \Sigma_p$ with line element

$$ds^2 = g_{AB}\mathrm{d}x^A\mathrm{d}x^B = \mathrm{e}^{2\alpha\rho}\tilde{g}_{\mu\nu}\mathrm{d}\tilde{x}^\mu\mathrm{d}\tilde{x}^\nu + \mathrm{e}^{2\beta\rho}\hat{g}_{ab}\mathrm{d}\hat{x}^a\mathrm{d}\hat{x}^b, \tag{9.2}$$

where α and β are constants, capital Latin indices refer to the D-dimensional spacetime, Greek indices refer to the external spacetime of dimension n with metric $\tilde{g}_{\mu\nu}$, lower case Latin indices refer to the internal manifold of dimension p with metric $\hat{g}_{ab}$, and ρ is a scalar that varies over the external spacetime only. H_k is a top-form that is either purely external ($k = n = D - p$) or purely internal ($k = p$) since there is no preference for one or the other at the level of the effective field theory. We will consider Σ_n to be either de Sitter dS_n or Minkowski $\mathbb{R}^{1,n-1}$, while Σ_p will be either a sphere S^p or a torus T^p.

We initially adopt a perturbative approach, working to leading order in the Gauss-Bonnet coupling, $f(\varphi)$. In this limit, we find that the MvRV bound obtained in [6] can be marginally violated and de Sitter solutions can still exist. This is true as long as a second parametric bound holds on the gradient of $f(\varphi)$. Unfortunately, this second bound does **not** hold for choices of $f(\varphi)$ and $g(\varphi)$ best motivated from string theory. To reinforce this latter point, we extend our analysis beyond perturbation theory, this time specialising to the string motivated choices for $f(\varphi)$ and $g(\varphi)$. Consistent with our perturbative analysis, we find that the MvRV bound is still a necessary condition for the existence of de Sitter solutions. However, it is no longer sufficient, suggesting that for string motivated potentials, the higher curvature corrections make it parametrically harder, not easier to find de Sitter vacua.

The structure of this chapter is as follows. In Sect. 9.1, we review the results of [6] and motivate the introduction of the Gauss-Bonnet term as a higher order curvature correction within the framework of heterotic strings. In Sect. 9.2, we gain some insight into the effects of the Gauss-Bonnet correction by performing a small coupling expansion. Then, in Sect. 9.3, we focus on specific string-inspired potentials and discuss in detail what conditions are necessary and sufficient for the existence of solutions. We conclude with a discussion in Sect.9.4.

9.1 Review of Flux Compactifications of the Heterotic String Action and Stability Conditions

In [6], the authors considered the existence and stability of $dS_n \times S^p$ flux compactifications. The action of [6] can be cast as Eq. (9.1) with $f(\varphi) = 0$ and with $k = p$, so that H_k is a top-form living in the internal space, providing the flux that aims to stabilise the potential for the scalar φ, and prevent the runaway. In this particular case, the coefficients α and β which parametrise the compactification are chosen to be

$$\alpha^2 = \frac{p}{2(n-2)(n+p-2)}, \qquad \beta = -\frac{n-2}{p}\alpha, \tag{9.3}$$

so that the n dimensional effective action obtained after dimensional reduction can be expressed in Einstein frame, with a canonical kinetic term for the breathing mode. The effective potential for φ and ρ in the external spacetime is found to be[2]

$$V_{\text{eff}} = -\frac{p(p-1)}{2\kappa_D^2} e^{2(\alpha-\beta)\rho} + Q^2 e^{2(n-1)\alpha\rho} g(\varphi) + V(\varphi) e^{2\alpha\rho} . \tag{9.4}$$

As a result, for constant φ and ρ, the metric field equations in the external n dimensional space take the form $\tilde{G}_{\mu\nu} = -V_{\text{eff}}\, \tilde{g}_{\mu\nu}$; Q is the magnetic charge associated with the internal top-form, fixed by the equations of motion to be

$$Q^2 = -e^{-2\alpha(n-1)\rho} \frac{V'}{g'} , \tag{9.5}$$

where the primes denote derivative with respect to φ. Demanding that the potential is positive at its minimum leads to the MvRV constraint

$$(p-1)\left|\frac{V'}{V}\right| \leq \left|\frac{g'}{g}\right| \qquad \text{or equivalently} \qquad (p-1)\left|\frac{V}{g}\right| \geq e^{-2\alpha(D-p-1)\rho} \frac{1}{Q^2} . \tag{9.6}$$

To check the stability of such solutions, one must specify the form of the scalar potential V and the coupling between the top-form and the potential g, while also stabilising the scalar. In [6], the authors consider a number of string-motivated scenarios with the key takeaway being the instability of the de Sitter minimum. For illustrative purposes, we present below the case where $V(\varphi) = V_0 e^{-\gamma\varphi}$ and $g(\varphi) = e^{\delta\varphi}$, for $\gamma, \delta > 0$. The Hessian of the potential along the φ, ρ directions is given by

$$\mathcal{H} = \frac{\partial^2 V_{\text{eff}}}{\partial\phi^i \partial\phi^j} , \qquad \phi^i = \{\varphi, \rho\} . \tag{9.7}$$

and its corresponding determinant by

$$\det(\mathcal{H}) = -\frac{(n-2)[\delta + \gamma(n-1)][\gamma(1-p)+\delta]\delta e^{2\delta\varphi + 2(n-1)\rho}}{4\gamma p} Q^4 . \tag{9.8}$$

Demanding that Eq. (9.6) holds at the minimum—or in other words $(p-1)\gamma \leq \delta$—fixes the determinant of the Hessian to be negative whenever $n > 2$. This means all the corresponding solutions are unstable. The only loophole to the above discussion is the case where the number of external dimensions $n = 2$. In this case, a specific analysis is required and one may obtain a meta-stable de Sitter solution while satisfying Eq. (9.6).

[2] Up to an overall internal volume factor which can be reabsorbed in a Weyl redefinition of the four-dimensional metric.

As mentioned in the introduction, here we expand upon the above obstruction by considering the effects of a Gauss-Bonnet term in the action Eq. (9.1). Let us now review the origin of this term. Our starting point is the effective action presented in [7]

$$S = \int \mathrm{d}^D x \sqrt{|g|} \left[\frac{R}{2\kappa_D^2} - \frac{1}{2}(\partial\varphi)^2 - V(\varphi) - g(\varphi)\,|H_3|^2 + f(\varphi)\mathcal{G} \right], \tag{9.9}$$

with $H_3 = \mathrm{d}B_2 + \omega_{\mathrm{L}} - \omega_{\mathrm{Y}}$ where B_2 is the Kalb-Ramond field, ω_{L} is the Lorentz spin connection, ω_{Y} is the Yang-Mills gauge connection, and with

$$f(\varphi) = \frac{\alpha'}{16\kappa_D^2} \mathrm{e}^{-\kappa_1\varphi}, \quad g(\varphi) = \frac{\mathrm{e}^{-2\kappa_1\varphi}}{\kappa_D^2}, \tag{9.10}$$

with $\kappa_1 = 2\kappa_D/\sqrt{D-2}$ and $D = 10$. The Gauss-Bonnet scalar is defined as

$$\mathcal{G} = R_{ABCD}R^{ABCD} - 4R_{AB}R^{AB} + R^2. \tag{9.11}$$

This is the combination of the quadratic curvature corrections that emerges in the effective theory for supersymmetric strings after super-completing the multiplet with the Lorentz-Chern-Simons terms, guaranteeing the absence of ghosts [8–10]. For $V = 0$, the action Eq. (9.9) describes the low-energy dynamics of the heterotic superstring [11]. Alternatively, by orbifolding $E_8 \times E_8$ strings one can obtain a non-supersymmetric theory whose low-energy limit is given by $SO(16) \times SO(16)$ strings, where the vacuum energy generates a potential [12],

$$V(\varphi) = \frac{0.037}{\alpha'^5} \mathrm{e}^{5\kappa_1\varphi}. \tag{9.12}$$

The effective action Eq. (9.9) is now a consistent truncation of this $SO(16) \times SO(16)$ action.

In this chapter, we shall work with a slight generalisation of Eq. (9.9), as given by Eq. (9.1), allowing arbitrary spacetime dimension, D, and a k-form that is not necessarily a 3-form. We will also consider more general potentials, focussing on exponentials with arbitrary coefficients and arbitrary slopes. Actually, these coefficients are not *entirely* arbitrary—we will assume they have the same sign as in the known example from the heterotic string Eq. (9.9). This means that all the cases of interest will follow the philosophy of [6], a runaway potential to be lifted by flux contributions. Note that the relative sign of the slopes is all that really matters in the corresponding potentials. This because one can always perform a redefinition of the scalar $\varphi \to -\varphi$. With our conventions, the runaway is assumed to occur in the limit $\varphi \to -\infty$, in contrast with the conventions of [6], but consistent with [7].

In the remaining sections, we consider compactified solutions of the action Eq. (9.1) that are a direct product of n-dimensional de Sitter/Minkowski times a p-sphere/torus. In terms of the Riemann tensors associated with the external and internal metrics, this translates as

$$\tilde{R}_{\mu\nu\rho\sigma} = \kappa \left(\tilde{g}_{\mu\rho} \tilde{g}_{\nu\sigma} - \tilde{g}_{\mu\sigma} \tilde{g}_{\nu\rho} \right), \tag{9.13}$$

$$\hat{R}_{abcd} = \lambda \left(\hat{g}_{ac} \hat{g}_{bd} - \hat{g}_{ad} \hat{g}_{bc} \right), \tag{9.14}$$

where κ is the curvature scale of the physical spacetime, while λ is the curvature scale of the internal spacetime ($\lambda > 0$ corresponds to a p-sphere and $\lambda = 0$ to a p-torus). We will also assume that k is either n or p, ensuring that the k-forms fills either the external or internal space with flux. We remark that in the standard heterotic case, with $D = 10$ and $k = 3$ this requires compactification on a 3-sphere or a 3-torus, down to seven external dimensions, requiring three more compact directions to recover a four dimensional universe. Of course, an explicit embedding of this kind with stable moduli is a non-trivial matter and further consistency considerations, such as tadpole cancellation [13], must be taken into account, but this is beyond the scope of this chapter.

Let us conclude by noting that the cases described in this chapter do not fall into the usual Maldacena-Nuñez no-go theorem [14] since the Gauss-Bonnet term represents higher curvature corrections to the action that are not considered in the original paper.

9.2 Extending the Montero-van Riet-Venken Bound in the Small Gauss-Bonnet Coupling Limit

The goal of this chapter is to study the impact of the higher curvature corrections on the MvRV bound for the existence of de Sitter solutions. To develop some intuition, we begin by studying the Gauss-Bonnet correction using a perturbative expansion in the dimensionless parameter $\varepsilon \equiv \kappa_D^4 \langle f(\varphi)/V(\varphi) \rangle \geq 0$, where $f(\varphi)$ and $V(\varphi)$ are evaluated at the vev of φ. As a result, we extend the MvRV bound Eq. (9.6) to account for the higher curvature terms and show that the condition on the existence of de Sitter vacua can be recast in the form

$$\left| \frac{g'}{g} \right| \geq (p-1) \left| \frac{V'}{V} \right| - \varepsilon \cdot h(\varphi, \rho) + \mathcal{O}\left(\varepsilon^2\right), \tag{9.15}$$

where the sign of $h(\varphi, \rho)$ determines whether, at linear order, the Gauss-Bonnet correction favours the existence of solutions or not. In particular, if $h > 0$, the bound is more easily satisfied and the higher curvature corrections should make it easier to find de Sitter solutions; if $h < 0$, the bound is harder to satisfy and de Sitter solutions are less forthcoming.

Throughout this section we consider the top form to be purely internal and we keep the sign of the slopes for the potential and the couplings as given in Eq. (9.10), without specifying the functional form:

$$\mathrm{sgn}\left(\frac{V'}{V}\right)=1\,,\quad \mathrm{sgn}\left(\frac{g'}{g}\right)=-1\,,\quad \mathrm{sgn}\left(\frac{f'}{f}\right)=-1\,. \tag{9.16}$$

The Lagrangian for the n-dimensional theory, after compactifying the D-dimensional action Eq. (9.1) in a p-dimensional internal space, is

$$\mathcal{L}=\sqrt{-\tilde{g}}\left\{\frac{\tilde{R}}{2\kappa_D^2}+\frac{\hat{R}}{2\kappa_D^2}\mathrm{e}^{2(\alpha-\beta)\rho}-\mathrm{e}^{2\alpha\rho}V(\varphi)-g(\varphi)Q^2\mathrm{e}^{2(n-1)\alpha\rho}\right.$$
$$\left.+f(\varphi)\left[\tilde{\mathcal{G}}\mathrm{e}^{-2\alpha\rho}+2\tilde{R}\hat{R}\mathrm{e}^{-2\beta\rho}+\hat{\mathcal{G}}\mathrm{e}^{2(\alpha-2\beta)\rho}\right]+\cdots\right\}\,, \tag{9.17}$$

where we have omitted the terms involving derivatives of the scalar fields, as we are interested in studying the system at the minimum in the (φ,ρ) directions. Extremising the Lagrangian with respect to the metric and the two fields yields the conditions,

$$Q^2=\frac{\mathrm{e}^{-2n\alpha\rho}}{g'(\varphi)}\left\{f'(\varphi)\left[\lambda^2p(p-1)(p-2)(p-3)\,\mathrm{e}^{4(\alpha-\beta)\rho}\right.\right.$$
$$\left.\left.+2\lambda\kappa n(n-1)p(p-1)\mathrm{e}^{2(\alpha-\beta)\rho}+\kappa^2n(n-1)(n-2)(n-3)\right]-\mathrm{e}^{4\alpha\rho}V'(\varphi)\right\}\,, \tag{9.18}$$

$$0=\lambda\left\{\frac{1}{\kappa_D^2}(p-1)(p+n-2)\,\mathrm{e}^{2(\alpha-\beta)\rho}+2(p-1)\,f(\varphi)\mathrm{e}^{-2\beta\rho}\left[2\kappa\,(n-2)\,n\,(n-1)\right.\right.$$
$$\left.\left.+\lambda\,(p-2)\,(p-3)\,(p+2n-4)\,\mathrm{e}^{2(\alpha-\beta)\rho}\right]\right\}-2Q^2g(\varphi)\,(n-1)\,\mathrm{e}^{2(n-1)\alpha\rho}$$
$$-2\kappa^2n\,(n-1)\,(n-2)\,(n-3)\,f(\varphi)\mathrm{e}^{-2\alpha\rho}-2V(\varphi)\mathrm{e}^{2\alpha\rho}\,, \tag{9.19}$$

$$0=\kappa\left\{2(n-1)(n-2)\,f(\varphi)\left[\kappa(n-3)(n-4)\mathrm{e}^{-2\alpha\rho}+2\lambda p(p-1)\mathrm{e}^{-2\beta\rho}\right]\right.$$
$$\left.+\frac{1}{\kappa_D^2}(n-1)(n-2)\right\}+\lambda p(p-1)\left[2\lambda(p-2)(p-3)\,f(\varphi)\mathrm{e}^{2(\alpha-2\beta)\rho}+\frac{\mathrm{e}^{2(\alpha-\beta)\rho}}{\kappa_D^2}\right]$$
$$-2\,V(\varphi)\mathrm{e}^{2\alpha\rho}-2\,g(\varphi)\,Q^2\mathrm{e}^{2(n-1)\alpha\rho}\,, \tag{9.20}$$

It is not possible to solve the above system for arbitrary potentials and dimensions. However, we can solve the system order by order in a perturbative Gauss-Bonnet coupling expansion.[3] To do so, we take the ansatz

$$Q^2 = Q_0^2 + \varepsilon\, Q_1^2\,, \quad \lambda = \lambda_0 + \varepsilon\,\lambda_1\,, \quad \kappa = \kappa_0 + \varepsilon\,\kappa_1\,, \tag{9.21}$$

where the zeroth order quantities solve the previous system of equations with $f(\phi) = 0$ and are given by

$$Q_0^2 = e^{-2(n-2)\alpha\rho}\left|\frac{V'}{g'}\right|\,, \tag{9.22}$$

$$\lambda_0 = \frac{2\kappa_D^2 V e^{2\beta\rho}}{(p-1)(n+p-2)}\left|\frac{g}{g'}\right|\left[\left|\frac{g'}{g}\right| + (n-1)\left|\frac{V'}{V}\right|\right]\,, \tag{9.23}$$

$$\kappa_0 = \frac{2\kappa_D^2 V e^{2\alpha\rho}}{(p+n-2)(n-1)}\left|\frac{g}{g'}\right|\left[\left|\frac{g'}{g}\right| - (p-1)\left|\frac{V'}{V}\right|\right]\,. \tag{9.24}$$

Solving for the ansatz Eq. (9.21) to leading order in the Gauss-Bonnet coupling, we find the external curvature to be

$$\kappa = \frac{2\kappa_D^2 V e^{2\alpha\rho}}{(p+n-2)(n-1)}\left|\frac{g}{g'}\right|\left[\left|\frac{g'}{g}\right| - (p-1)\left|\frac{V'}{V}\right| + \varepsilon h_{n,p} + \mathcal{O}\left(\varepsilon^2\right)\right]\,, \tag{9.25}$$

where (φ, ρ) are understood to be evaluated at their vevs. Demanding that the external curvature is positive leads to an extended condition of the form Eq. (9.15), with the leading order correction given by a polynomial in $|g/g'|$

$$\begin{aligned} h_{n,p} = \frac{V^2}{(p-1)(n-1)(p+n-2)} &\left[A_{n,p}\left|\frac{g'}{g}\right| + B_{n,p}\left|\frac{f'}{f}\right| + C_{n,p}\left|\frac{V'}{V}\right|\right. \\ &\left. + \left(D_{n,p}\left|\frac{f'}{f}\right| + E_{n,p}\left|\frac{V'}{V}\right|\right)\left|\frac{V'}{V}\right|\left|\frac{g}{g'}\right| + F_{n,p}\left|\frac{f'}{f}\right|\left|\frac{V'}{V}\right|^2\left|\frac{g}{g'}\right|^2\right]\,. \end{aligned} \tag{9.26}$$

The coefficients of the polynomial depend on the dimensions (n, p):

$$A_{n,p} = 4n(p-7)(1-p) + 4n^2(1-p) + 4p(p-11) + 48\,, \tag{9.27}$$

$$B_{n,p} = -4\left[(n-1)p^2 + (n^2-4n+3)p - n^2 + 3n\right](p-1)\,, \tag{9.28}$$

$$C_{n,p} = 8\left[(n-1)p^3 + (n^2-11n+14)p^2 - (2n^2-19n+27)p + n^2 - 7n + 12\right]\,, \tag{9.29}$$

$$D_{n,p} = 8(p-n)(np-n-p+3)(p-1)\,, \tag{9.30}$$

$$E_{n,p} = 4\left[2(3n-5)p^3 - (2n^2+15n-31)p^2 + (3n^2+12n-31)p + n^2 - 7n + 12\right]\,, \tag{9.31}$$

[3] Note that Gauss-Bonnet gravity, being quadratic in curvature, typically yields two distinct vacua, one which is perturbative in the coupling, f, and one which is non-perturbative. Our analysis here does not capture the latter, which one might have expected to be unstable anyway [15].

$$F_{n,p} = 4(p-1)(4n^2p^2 - 7n^2p - 7np^2 + n^2 + 14np + p^2 - 3n - 3p)\,. \tag{9.32}$$

Although these expressions for the coefficients are not very illuminating, we can remark on some interesting behaviour.

For the particular stringy potentials given in Eq. (9.10) and with the well motivated choice, $n = 4$, $p = 6$ we find that $h_{n,p} > 0$. However, this does not point to the existence of new de Sitter solutions as the MvRV bound is already strongly violated at leading order. For the Gauss-Bonnet corrections to yield something interesting in this perturbative set-up, our best hope is ask what happens when the MkRV bound is *only just* violated at leading order. In particular, we consider the case where

$$\left|\frac{g'}{g}\right| = (p-1)\left|\frac{V'}{V}\right|(1-\Delta) \tag{9.33}$$

for some $0 < \Delta \ll \varepsilon$. In such a scenario, the existence of de Sitter solutions is only marginally ruled out by the original MvRV bound Eq. (9.6). Could the curvature corrections rule them back in? In this particular case we find that

$$h_{n,p} = \frac{4p(p-2)(p-3)}{(p-1)^3}V^2\left[\left|\frac{g'}{g}\right| - (p-1)\left|\frac{f'}{f}\right|\right][1+\mathcal{O}(\Delta)]\,, \tag{9.34}$$

Since $0 < \Delta \ll \varepsilon$, the external curvature is dominated by this form of $h_{n,p}$, giving

$$\kappa \approx \frac{8\kappa_D^2 p(p-2)(p-3)V^3 e^{2\alpha\rho}}{(p+n-2)(n-1)(p-1)^3}\varepsilon\left|\frac{g}{g'}\right|\left[\left|\frac{g'}{g}\right| - (p-1)\left|\frac{f'}{f}\right|\right] \tag{9.35}$$

which is clearly positive whenever

$$\left|\frac{g'}{g}\right| > (p-1)\left|\frac{f'}{f}\right|\,. \tag{9.36}$$

This suggest that de Sitter solutions *might* be possible for these parametric choices, although we are not aware of any stringy motivated supergravity model that satisfies them.

Another possibility is to consider the case where $|g'/g| \ll |V'/V|$. When this hierarchy is big enough, the last term in Eq. (9.26) could be the dominant contribution to the external curvature, dominating over the zeroth order piece in the perturbative expansion in the Gauss-Bonnet coupling. Whilst this casts some doubt on the validity of the expansion, it would, if true, suggest that

$$\kappa \approx \frac{2\kappa_D^2 V^3 e^{2\alpha\rho}}{(p+n-2)^2(n-1)^2(p-1)}\varepsilon F_{n,p}\left|\frac{f'}{f}\right|\left|\frac{V'}{V}\right|^2\left|\frac{g}{g'}\right|^3 \tag{9.37}$$

This could point towards new de Sitter solutions since we also have that $F_{n,p} > 0$ for $n, p \geq 2$. Of course, such an extreme hierarchy is not especially well motivated since generically we expect $|g'/g|$ and $|V'/V|$ to be $\mathcal{O}(1)$ as both couplings are expected to be fixed by Weyl rescaling to Einstein frame.

9.3 Non-perturbative Conditions for the Existence of de Sitter Solutions with a Gauss-Bonnet Term

We shall now establish some exact results, seeking de Sitter and non-trivial Minkowski vacua beyond the perturbative approach of the previous section. To this end, it is necessary to fix some of the potentials in order to make concrete progress. We choose the Gauss-Bonnet potential f and the gauge coupling g to coincide with the heterotic string, as per Eq. (9.10), so that

$$f(\varphi) = \frac{\alpha'}{16\kappa_D^2} e^{-\kappa_1 \varphi}, \quad g(\varphi) = \frac{e^{-2\kappa_1 \varphi}}{\kappa_D^2} \tag{9.38}$$

while the remaining potential is given more generally as

$$V(\varphi) = V_0 e^{q\kappa_1 \varphi}, \tag{9.39}$$

with $V_0 > 0$ and $q > 0$ a free dimensionless parameter. There is sufficient generality left in V to mirror the analysis of [6], and investigate parametric constraints on the slope of $\ln V$. Note that for the string motivated scenario of [12], we have $q = 5$ and $V_0 = \frac{0.037}{\alpha'^5}$.

In [6], the slope of $\ln V$ is constrained in the form of the MvRV bound Eq. (9.6). For these exponential potentials, this bound takes a particularly simple form

$$q \leq \frac{2}{p-1}. \tag{9.40}$$

Our goal here is to ask if the parametric constraint on the slope of $\ln V$ is affected by the Gauss-Bonnet correction. We will also allow for both internal and external fluxes, further generalising [6]. If H_k is allowed to carry external flux, we find that there are only trivial vacua: a Minkowski external space, vanishing flux and a toroidal internal manifold. If, instead, H_k is allowed to carry internal flux, the set-up is more interesting. The perturbative analysis of the previous section suggested that one could find de Sitter solutions even when the MvRV bound is violated, provided an extended condition Eq. (9.36) holds for the Gauss-Bonnet potentials. However, it turns out that this extended bound is *not* satisfied for the string motivated potentials for f and g given by Eq. (9.38). Indeed, we shall see that in this instance, the MvRV condition Eq. (9.40) is necessary for the existence of de Sitter solutions, but it is not sufficient. This means the higher curvature correction makes de Sitter solutions no more forthcoming, constraining the parametric dependence of the scalar potential $\ln V$ at least as much as in [6].

Note that in this section, we shall not assume the particular choice of α and β given in Eq. (9.3), but will keep things general. Furthermore, we work directly with the D-dimensional field equations derived from the variation of the action Eq. (9.1), which are given by

$$\begin{aligned}
0 = f(\varphi)&\left(4R_A^{\ C}R_{BC} + \frac{1}{2}\mathcal{G}g_{AB} - 2RR_{AB} + 4R^{CD}R_{ACBD} - 2R_A^{\ CDE}R_{BCDE}\right)\\
&+\frac{1}{2}\nabla_A\varphi\nabla_B\varphi - \frac{1}{4}(\nabla\varphi)^2 g_{AB} + 2f'(\varphi)\Big(R\nabla_A\nabla_B\varphi - 4R_{(B}^{\ C}\nabla_{A)}\nabla_C\varphi + 2G_{AB}\Box\varphi\\
&+2g_{AB}R^{CD}\nabla_C\nabla_D\varphi - 2R_{ACBD}\nabla^C\nabla^D\varphi\Big) + 2f''(\varphi)\,[R\nabla_A\varphi\nabla_B\varphi\\
&-4R_{C(B}\nabla_{A)}\varphi\nabla^C\varphi + 2G_{AB}(\nabla\varphi)^2 + 2g_{AB}R^{CD}\nabla_B\varphi\nabla_C\varphi - 2R_{ACBD}\nabla^C\varphi\nabla^D\varphi\Big]\\
&-\frac{1}{2\kappa_D^2}G_{AB} - g(\varphi)\left[\frac{1}{2k!}H_{A_1\ldots A_k}H^{A_1\ldots A_k}g_{AB} - \frac{1}{(k-1)!}H_{A_1\ldots A_{k-1}A}H^{A_1\ldots A_{k-1}}{}_B\right]\\
&-\frac{1}{2}V(\varphi)g_{AB},
\end{aligned} \tag{9.41}$$

$$0 = \Box\varphi + f'(\varphi)\mathcal{G} - V'(\varphi) - \frac{g'(\varphi)}{k!}H_{A_1\ldots A_k}H^{A_1\ldots A_k}, \tag{9.42}$$

$$0 = \nabla_B\left[g(\varphi)H^{A_1\ldots A_{k-1}B}\right]. \tag{9.43}$$

9.3.1 Internal Flux

If we assume that H is a top form along the internal manifold, then the solution to Eq. (9.43) is given by

$$H_{a_1\ldots a_p} = Q\hat{\epsilon}_{a_1\ldots a_p} \tag{9.44}$$

where Q is a constant, all other components of H vanish, and $\hat{\epsilon}_{a_1\ldots a_p}$ is the volume form associated with the internal submanifold ($\hat{\epsilon}_{1\ldots p} = \sqrt{\hat{g}}$, etc.). We substitute this solution, together with Eqs. (9.13) and (9.14) into Eqs. (9.41) and (9.42), projecting equation (9.41) over $\tilde{g}_{\mu\nu}$ and $\hat{g}_{ab}$ respectively. Let us define the following master variables

$$\begin{aligned}
K &= \kappa e^{2(\beta-\alpha)\rho}, & \hat{H} &= Q^2 e^{-2\kappa_1\varphi - 2(p-1)\beta\rho}, & U &= V_0\kappa_D^2 e^{q\kappa_1\varphi + 2\beta\rho},\\
X_1 &= \frac{\alpha'}{16}e^{-\kappa_1\varphi - 2\beta\rho}, & X_2 &= \frac{\alpha'}{16}\kappa e^{-\kappa_1\varphi - 2\alpha\rho}, & X_3 &= \frac{\alpha'}{16}\kappa^2 e^{-\kappa_1\varphi + 2(\beta-2\alpha)\rho},
\end{aligned} \tag{9.45}$$

Given that $\alpha' > 0$, for a consistent de Sitter (or Minkowski) vacuum, we necessarily have that all of the master variables, with the possible exception of U, are non-negative. This will be crucial to our subsequent proofs. The system of equations can now be cast in the following linear form, which is particularly useful for analysing the system

$$\begin{aligned} 0 = -2\hat{H} - 2U + (n-1)(n-2)[K + 2(n-3)(n-4)X_3] \\ + \lambda p(p-1)[1 + 4(n-1)(n-2)X_2] + 2\lambda^2 p(p-1)(p-2)(p-3)X_1, \end{aligned} \tag{9.46}$$

$$\begin{aligned} 0 = 2\hat{H} - 2U + n(n-1)[K + 2(n-2)(n-3)X_3] \\ + \lambda(p-1)(p-2)[1 + 4n(n-1)X_2] + 2\lambda^2(p-1)(p-2)(p-3)(p-4)X_1, \end{aligned} \tag{9.47}$$

$$\begin{aligned} 0 = 2\hat{H} - qU - n(n-1)(n-2)(n-3)X_3 - 2\lambda n(n-1)p(p-1)X_2 \\ - \lambda^2 p(p-1)(p-2)(p-3)X_1. \end{aligned} \tag{9.48}$$

Given that $n \geq 2$ and $q > 0$, we can always solve the above equations for U, $\hat{H}$ and K in terms of X_1, X_2 and X_3. In particular, we have

$$\begin{aligned} K = K_{\rm sol}^{(\rm int)}(X_1, X_2, X_3) \equiv -\frac{1}{(n-1)[2+(n-1)q]} \{\lambda(p-1)[q(p-1)-2] \\ + 2\lambda^2(p-1)(p-2)(p-3)[p-4+(p-2)q]X_1 \\ + 4\lambda(p-1)(n-1)\{-p(q-2) + n[p-2+(p-1)q]\}X_2 \\ + 2(n-1)(n-2)(n-3)[4+n+(n-2)q]X_3\}. \end{aligned} \tag{9.49}$$

The idea behind the proof is to show that K is maximal at $(X_1, X_2, X_3) = (0, 0, 0)$ (restricting of course to non-negative X_i values), but that even at its maximum, $K < 0$ as soon as we choose $q > 2/(p-1)$. Since K carries the same sign as the external curvature, this would prove the absence of de Sitter solutions whenever $q > 2/(p-1)$. Said another way, the MvRV condition Eq. (9.40) is a necessary condition for the existence of de Sitter vacua, even in the presence of the Gauss-Bonnet correction.

To this end, we consider

$$\frac{\partial K_{\rm sol}^{(\rm int)}}{\partial X_1} = -\frac{2\lambda^2(p-1)(p-2)(p-3)[p-4+(p-2)q]}{(n-1)[2+(n-1)q]}, \tag{9.50}$$

$$\frac{\partial K_{\rm sol}^{(\rm int)}}{\partial X_2} = -\frac{4\lambda(p-1)\{-p(q-2)+n[p-2+(p-1)q]\}}{2+(n-1)q}, \tag{9.51}$$

$$\frac{\partial K_{\rm sol}^{(\rm int)}}{\partial X_3} = -\frac{2(n-2)(n-3)[n+4+(n-2)q]}{2+(n-1)q}. \tag{9.52}$$

$\partial K_{\text{sol}}^{(\text{int})}/\partial X_3$ is obviously negative. For $p = 1, 2, 3$, $\partial K_{\text{sol}}^{(\text{int})}/\partial X_1 = 0$ while for $p \geq 4$, $\partial K_{\text{sol}}^{(\text{int})}/\partial X_1$ is also obviously negative. Finally, for $p = 1$, $\partial K_{\text{sol}}^{(\text{int})}/\partial X_2 = 0$, while for $p \geq 2$ and $n \geq 2$,

$$-p(q-2) + n[p - 2 + (p-1)q] \geq (p-2)q + 4(p-1) > 0. \tag{9.53}$$

Hence, $\partial K_{\text{sol}}^{(\text{int})}/\partial X_2 \leq 0$ in all cases, and $K_{\text{sol}}^{(\text{int})}$ is indeed maximal at $(X_1, X_2, X_3) = (0, 0, 0)$, where it takes the value

$$K_{\text{max}}^{(\text{int})} \equiv K_{\text{sol}}^{(\text{int})}(0, 0, 0) = \frac{\lambda(p-1)[2 - (p-1)q]}{(n-1)[2 + (n-1)q]}. \tag{9.54}$$

The above equation shows that K can vanish for $\lambda = 0$ (p-torus) or $p = 1$ (1-sphere), i.e. when the curvature of the internal manifold vanishes. These solutions correspond to a Minkowski spacetime, and exist only for a vanishing flux Q and a vanishing potential $V(\varphi)$, which makes them uninteresting. The existence of a de Sitter solution, on the other hand, requires that $K > 0$. This is not possible if $q > 2/(p-1)$, as shown by Eq. (9.54), which completes our proof. The condition Eq. (9.40) is necessary for de Sitter solutions to exist.

In terms of the variables K, $\hat{H}$, U, X_i, Eq. (9.40) is also sufficient; indeed, it is possible to show that, when $q \leq 2/(p-1)$, K, $\hat{H}$ and U are all positive in the neighbourhood of the point $(X_1, X_2, X_3) = (0, 0, 0)$. However, this is an artifact of the choice of variables. Sending all X_i to 0 actually corresponds to sending the Gauss-Bonnet coupling to 0. Hence, it is not surprising to find that condition Eq. (9.40) is necessary and sufficient in this case, as we are back to the action studied in [6]. If instead, we think of the Gauss-Bonnet coupling as fixed, we can show that the condition Eq. (9.40) is not sufficient, by an analysis that is very similar to the one we carried out in Sect. 9.2. Indeed, let us saturate the bound Eq. (9.40) by choosing $q = 2/(p-1)$ — which corresponds to being on the edge of the existence region when the Gauss-Bonnet coupling is absent. Then, $K_{\text{sol}}^{(\text{int})}$ becomes

$$\begin{aligned} K_{\text{sol}}^{(\text{int})} = -\frac{1}{(n-1)(n+p-2)} \Big\{ &(n-1)(n-2)(n-3)[n - 8 + (4+n)p]X_2 \\ &+ 2p(p-1)(n-1)[(2+n)p - n - 4]\lambda X_3 + p(p-1)(p-2)(p-3)^2\lambda^2 X_1 \Big\}. \end{aligned} \tag{9.55}$$

It is easy to check that for $p \geq 2$ (we already saw that the case $p = 1$ was uninteresting), the coefficients of X_1, X_2 and X_3 in the above equation are all negative. The coefficient of X_3 is actually strictly negative. Hence, there can be no de Sitter (or Minkowski) solution for this value of q—and neighbouring ones, by continuity. Therefore, Eq. (9.40) is not sufficient. This reinforces the conclusion of Sect. 9.2, as it is a non-perturbative statement. The presence of the Gauss-Bonnet term never enlarges the space of solutions, but instead reduces it, at least when the potentials, f and g, take on the form best motivated by string theory.

9.3.2 External Flux

So far, we have assumed H to be an internal flux, and, as we have demonstrated, this leads to the absence of de Sitter solutions to the effective action Eq. (9.1) as long as $q > 2/(p-1)$. Now let us consider the opposite scenario: that H is a top form along the external directions. In this case, the solution to Eq. (9.43) is given by

$$H_{\mu_1\ldots\mu_n} = \frac{Q}{g(\varphi)} e^{(n\alpha - p\beta)\rho} \tilde{\epsilon}_{\mu_1\ldots\mu_n} \tag{9.56}$$

where Q is a constant, all other components of H vanish, and $\tilde{\epsilon}_{\mu_1\ldots\mu_n}$ is the volume form associated with the internal submanifold ($\tilde{\epsilon}_{1\ldots n} = \sqrt{-\tilde{g}}$, etc.). We will prove that no de Sitter solutions at all exist in this case, for any value of q. The proof is very similar to the one presented in Sect. 9.3.1. Employing the same variables U, K, X_i as defined in Eq. (9.45), and trading $\hat{H}$ for

$$\tilde{H} = Q^2 e^{2\kappa_1\varphi - 2(p-1)\beta\rho}, \tag{9.57}$$

the field equations are identical to Eqs. (9.46)–(9.48) upon the substitution $\hat{H} \to \tilde{H}$ in Eqs. (9.46)–(9.47) and $\hat{H} \to -\tilde{H}$ in Eq. (9.48). We need to distinguish a few more sub-cases with respect to the internal case, depending on how q compares to $2/(n-1)$.

$q > 2/(n-1)$

In this case, the proof is almost exactly identical to the internal case. We solve the field equations for K, U and $\tilde{H}$ in terms of X_1, X_2 and X_3, to obtain

$$\begin{aligned} K = K_{\text{sol}}^{(\text{ext})}(X_1, X_2, X_3) \equiv -\frac{1}{(n-1)[(n-1)q-2]} &\{\lambda(p-1)[q(p-1)+2] \\ &+ 2\lambda^2(p-1)(p-2)(p-3)[p+4+(p-2)q]X_1 \\ &+ 4\lambda(n-1)(p-1)\{-p(q+2)+n[p+2+(p-1)q]\}X_2 \\ &+ 2(n-1)(n-2)(n-3)[n-4+(n-2)q]X_3\}\,. \end{aligned} \tag{9.58}$$

It immediately follows that

$$\frac{\partial K_{\text{sol}}^{(\text{ext})}}{\partial X_1} = -\frac{2\lambda^2(p-1)(p-2)(p-3)[p+4+(p-2)q]}{(n-1)[(n-1)q-2]}, \tag{9.59}$$

$$\frac{\partial K_{\text{sol}}^{(\text{ext})}}{\partial X_2} = -\frac{4\lambda(p-1)\{-p(q+2)+n[p+2+(p-1)q]\}}{(n-1)q-2}, \tag{9.60}$$

$$\frac{\partial K_{\text{sol}}^{(\text{ext})}}{\partial X_3} = -\frac{2(n-2)(n-3)[n-4+(n-2)q]}{(n-1)q-2}. \tag{9.61}$$

Since $q > 2/(n-1)$, $\partial K_{\mathrm{sol}}^{(\mathrm{ext})}/\partial X_1$ is obviously negative. For $n = 2, 3$, $\partial K_{\mathrm{sol}}^{(\mathrm{ext})}/\partial X_3 = 0$ while for $n \geq 4$, $\partial K_{\mathrm{sol}}^{(\mathrm{ext})}/\partial X_3$ is also obviously negative. Finally, for $p = 1$, $\partial K_{\mathrm{sol}}^{(\mathrm{ext})}/\partial X_2 = 0$, while for $p \geq 2$ and $n \geq 2$,

$$-p(q+2) + n[p+2+(p-1)q] \geq (p-2)q + 4 > 0. \tag{9.62}$$

Hence, $\partial K_{\mathrm{sol}}^{(\mathrm{ext})}/\partial X_2 \leq 0$ in all cases, and $K_{\mathrm{sol}}^{(\mathrm{ext})}$ is maximal at $(X_1, X_2, X_3) = (0, 0, 0)$, where it takes the value

$$K_{\max}^{(\mathrm{ext})} \equiv K_{\mathrm{sol}}^{(\mathrm{ext})}(0,0,0) = -\frac{\lambda(p-1)[2+(p-1)q]}{(n-1)[(n-1)q-2]}. \tag{9.63}$$

This is always negative in the sub-case that we are considering, hence no de Sitter solutions exist. Again, trivial Minkowski solutions can exist for $p = 1$ or $\lambda = 0$.
$q < 2/(n-1)$
Once again, the proof is very analogous to the previous ones, but now we consider the solution for $\tilde{H}$ rather than K. This is given by

$$\begin{aligned}\tilde{H} = \tilde{H}_{\mathrm{sol}}(X_1, X_2, X_3) \equiv \frac{1}{2[(n-1)q-2]}&\{\lambda(p-1)q(n+p-2)\\ &+ 2\lambda^2(p-1)(p-2)(p-3)[p+(2n+p-4)q]X_1\\ &+ 4\lambda n(n-1)(p-1)[p+(n-2)q]X_2\\ &-2n(n-1)(n-2)(n-3)(q-1)X_3\}.\end{aligned} \tag{9.64}$$

and therefore,

$$\frac{\partial \tilde{H}_{\mathrm{sol}}}{\partial X_1} = \frac{2\lambda^2(p-1)(p-2)(p-3)[p+(p+2n-4)q]}{(n-1)[(n-1)q-2]}, \tag{9.65}$$

$$\frac{\partial \tilde{H}_{\mathrm{sol}}}{\partial X_2} = \frac{4\lambda n(n-1)(p-1)[p+(n-2)q]}{(n-1)q-2}, \tag{9.66}$$

$$\frac{\partial \tilde{H}_{\mathrm{sol}}}{\partial X_3} = \frac{n(n-1)(n-2)(n-3)(1-q)}{(n-1)q-2}. \tag{9.67}$$

$\partial \tilde{H}_{\mathrm{sol}}/\partial X_1$ and $\partial \tilde{H}_{\mathrm{sol}}/\partial X_2$ are obviously negative in the subcase we consider. For $n \leq 3$, $\partial \tilde{H}_{\mathrm{sol}}/\partial X_3 = 0$. For $n \geq 4$, since $q < 2/(n-1)$, $1 - q > 0$ and therefore $\partial \tilde{H}_{\mathrm{sol}}/\partial X_3$ is negative. Hence, $\tilde{H}_{\mathrm{sol}}$ is maximal at $(X_1, X_2, X_3) = (0, 0, 0)$, where it takes the value

$$\tilde{H}_{\max} \equiv \tilde{H}_{\mathrm{sol}}(0,0,0) = \frac{\lambda q(p-1)(n+p-2)}{2[(n-1)q-2]} < 0. \tag{9.68}$$

Since $\tilde{H}$ should be non-negative for real values of the flux, we conclude that the underlying assumptions are inconsistent and therefore no de Sitter or non-trivial Minkowski solutions can exist.

$q = 2/(n-1)$

In the case where q takes exactly the value $2/(n-1)$, we can no longer solve the system of equations for K, U and $\tilde{H}$ simultaneously. However, we can solve it for X_2, U and $\tilde{H}$ instead, provided that $\lambda \neq 0$ and $p \geq 2$ (if this is not the case, once again only trivial Minkowski solutions can exist). We obtain

$$\begin{aligned} X_2 = X_{2,\,\mathrm{sol}}(X_1, X_3) \equiv &-\frac{1}{2\lambda n(n-1)(p-1)[(n-1)p+2(n-2)]} \\ &\left\{\lambda^2(p-1)(p-2)(p-3)[p(n+1)+4(n-2)]X_1 \right. \\ &\left. +n(n-1)(n-2)(n-3)^2 X_3 + \lambda(p-1)(n+p-2)\right\}. \end{aligned} \tag{9.69}$$

Hence,

$$\frac{\partial \tilde{X}_{2,\,\mathrm{sol}}}{\partial X_1} = -\frac{\lambda(p-2)(p-3)[p(n+1)+4(n-2)]}{2n(n-1)[(n-1)p+2(n-2)]}, \tag{9.70}$$

$$\frac{\partial \tilde{X}_{2,\,\mathrm{sol}}}{\partial X_3} = -\frac{(n-2)(n-3)^2}{2\lambda(p-1)[(n-1)p+2(n-2)]}, \tag{9.71}$$

These quantities are obviously negative. Therefore, $X_{2,\,\mathrm{sol}}$ is maximal at $(X_1, X_3) = (0, 0)$, where it takes the value

$$X_{2,\,\mathrm{max}} \equiv \tilde{X}_{2,\,\mathrm{sol}}(0,0) = -\frac{n+p-2}{2n(n-1)[(n-1)p+2(n-2)]} < 0. \tag{9.72}$$

Again, for consistent de Sitter or non-trivial Minkowski solultions, X_2 should be non-negative, in contradiction with the above inequality. This concludes the proof for the external case.

9.4 Discussion

We have considered solutions to a general class of gravitational actions, motivated by quadratic curvature corrections to the effective action for the heterotic string. The actions include a scalar field (identified with the dilaton), a k-form field strength, and gravity. We considered warped compactifications on spheres and tori, with k-form flux and investigated the existence of de Sitter solutions along the external directions. A similar analysis was carried out in [6] without the quadratic curvature corrections, which in our case corresponded to the Gauss-Bonnet operator coupled to the dilaton. In [6], the MvRV bound Eq. (9.6) was derived for the dilaton potentials, specifying the condition for de Sitter solutions to exist. In the presence of the quadratic curvature corrections, we saw that this bound could be violated perturbatively and de Sitter solutions could still exist, as long as a second bound Eq. (9.36) on the Gauss-Bonnet coupling also held. However, for potentials that are best motivated by string theory, the second bound does not hold, suggesting no violation of the MvRV bound is possible in this instance. Indeed, for these well motivated set-ups, we were also able to show that the MvRV bound was a necessary, but not sufficient, condition for the existence of de Sitter solutions in the presence of quadratic curvature corrections. This leads to our main conclusion: higher curvature corrections that are well motivated by string theory generically made it parametrically harder, not easier, to find de Sitter solutions.

It would be very interesting to ask if this were a generic feature of higher order corrections in all versions of string theory. Clearly this is difficult to establish and more evidence needs to be found. For type II strings, curvature corrections kick in at fourth order [16], while for heterotic strings we may also consider higher order mixing between the gauge field and curvature [11].

We can make a crude and simplistic argument to support a claim that curvature corrections are unlikely to improve the search for de Sitter vacua, at least at finite order in the curvature expansion. In D dimensional General Relativity without sources, the absence of de Sitter vacua can roughly be understood in terms of an energetic balance between the curvature of the external space and that of the compact space. If the compact space carries positive or vanishing curvature (as in a sphere or a torus), then the external space looks to counter that, delivering negative or vanishing curvature (as in AdS or Minkowski). For higher curvature operators to help facilitate new de Sitter vacua, the details of this balance have to change. In particular, on one side of the energetic balance, a positive curvature space (be it a sphere along the compact directions, or de Sitter along the external directions) must contribute negative energy through the higher order operators. This might indicate the presence of ghost-like instabilities and the associated negative energy excitations. Such instabilities should certainly be absent in effective actions derived from string theory.

Of course, the situation becomes more complicated in the presence of additional fields and sources, including branes, fluxes, Casimir energies, and with more exotic compact spaces. The challenge here is to retain calculational control over the effective description, be it parametrically or even just numerically. Indeed, the level of control one is willing to sacrifice seems to be at the core of the ongoing debate over de Sitter space and the *Swampland*.

References

1. Dudas E, Ferrara S, Kehagias A, Sagnotti A (2015) Properties of nilpotent supergravity. JHEP 09:217. https://doi.org/10.1007/JHEP09(2015)217, arxiv:https://arxiv.org/abs/1507.07842
2. Bergshoeff EA, Freedman DZ, Kallosh R, Van Proeyen A (2015) Pure de Sitter supergravity. Phys Rev D 92:085040. https://doi.org/10.1103/PhysRevD.93.069901, arxiv:https://arxiv.org/abs/1507.08264

3. Bandos I, Martucci L, Sorokin D, Tonin M (2016) Brane induced supersymmetry breaking and de Sitter supergravity. JHEP 02:080. https://doi.org/10.1007/JHEP02(2016)080, arxiv:https://arxiv.org/abs/1511.03024]
4. Nagy S, Padilla A, Zavala I (2020) The super-Stückelberg procedure and dS in pure supergravity. Proc R Soc Lond A 476:20200035 https://doi.org/10.1098/rspa.2020.0035, arxiv:https://arxiv.org/abs/1910.14349
5. Bansal S, Nagy S, Padilla A, Zavala I (2021) Unimodular vs nilpotent superfield approach to pure dS supergravity. JHEP 01:146. https://doi.org/10.1007/JHEP01(2021)146, arxiv:https://arxiv.org/abs/2010.13758
6. Montero M, Van Riet T, Venken G (2020) A dS obstruction and its phenomenological consequences. JHEP 05:114. https://doi.org/10.1007/JHEP05(2020)114, arxiv:https://arxiv.org/abs/2001.11023
7. Campbell BA, Duncan MJ, Kaloper N, Olive KA (1991) Gravitational dynamics with Lorentz Chern-Simons terms. Nucl Phys B 351:778. https://doi.org/10.1016/S0550-3213(05)80045-8
8. Zwiebach B (1985) Curvature squared terms and string theories. Phys Lett B 156:315. https://doi.org/10.1016/0370-2693(85)91616-8
9. Zumino B (1986) Gravity theories in more than four-dimensions. Phys Rept 137:109
10. Cecotti S, Ferrara S, Girardello L, Porrati M (1985) Lorentz Chern-simons Terms in N=1 four-dimensional supergravity consistent with supersymmetry and string compactification. Phys Lett B 164:46. https://doi.org/10.1016/0370-2693(85)90028-0
11. Gross DJ, Sloan JH (1987) The quartic effective action for the heterotic string. Nucl Phys B 291(1987)41. https://doi.org/10.1016/0550-3213(87)90465-2
12. Alvarez-Gaume L, Ginsparg PH, Moore GW, Vafa C (1986) An O(16) x O(16) heterotic string. Phys Lett B 171:155. https://doi.org/10.1016/0370-2693(86)91524-8
13. Becker K, Tseng L-S (2006) Heterotic flux compactifications and their moduli. Nucl Phys B 741:162. https://doi.org/10.1016/j.nuclphysb.2006.02.013, arxiv:https://arxiv.org/abs/hep-th/0509131hep-th/0509131
14. Maldacena JM, Nunez C (2001) Supergravity description of field theories on curved manifolds and a no go theorem. Int J Mod Phys A 16:822. https://doi.org/10.1142/S0217751X01003937, arxiv:https://arxiv.org/abs/hep-th/0007018hep-th/0007018
15. Charmousis C, Padilla A (2008) The instability of Vacua in Gauss-Bonnet gravity. JHEP 12:038. https://doi.org/10.1088/1126-6708/2008/12/038, arxiv:https://arxiv.org/abs/0807.28640
16. Gross DJ, Witten E (1986) Superstring modifications of Einstein's equations. Nucl Phys B 277:1. https://doi.org/10.1016/0550-3213(86)90429-3

Chapter 10
Conclusions and Outlooks

In this thesis, we began by developing the necessary mathematical machinery to understand Calabi-Yau compactifications and proceeded to study the perturbative tree-level type IIB supergravity action in Chap. 2. We saw that the Kähler sector remained flat and anticipated that to be able to generate a potential for the Kähler moduli we would need to introduce perturbative and non-perturbative corrections to the tree-level action. This prompted the introduction of the Swampland programme, in particular with reference to the phenomenological side of the Swampland, and the questions of parametric control (or lack thereof).

In Chap. 3, we introduced a number of important open questions in modern cosmology thereby focusing the aim of the remainder of the thesis on obtaining explanations for some of these questions within the framework of string theory. Of particular interest were the naturalness and hierarchy problem, as these will plague any effective description of a "theory of everything",[1] and the nature of dark energy.

The last two chapters of the first half of the thesis explore these aspects of modern cosmology from the point of view of effective descriptions motivated by supergravity. In Chap. 4 we introduce the clockwork mechanism which aims to naturally produce scale hierarchies in theories with many scalars. In the original formulation, a theory with N scalars obtains a massless direction and $N - 1$ heavy fields. Furthermore, we apply this mechanism to a cosmological setting by introducing the Dvali-Kaloper-Sorbo 4-form mixing. This mixing can softly break the clockwork symmetry, giving a small mass to the originally massless clockwork mode while preserving the hierarchy between this new light mode and the heavy modes. We anticipated that for some $\mathcal{O}(100)$ scalar fields the light clockwork mass would be of order the dark energy scale observed today and thus this mechanism could prove useful to naturally generate dark

[1] On this occasion, by "theory of everything" we refer to any theory encompassing many different physical scales, as would be expected from a UV complete description of our Universe arising from $\mathcal{O}(\mp)$ physics.

F. Cunillera, *Dark Energy*, Springer Theses,
https://doi.org/10.1007/978-3-031-21468-4_10

energy models. We then provided a no-go theorem for embedding this mechanism into the tree-level type IIA supergravity.

In ongoing research, the question remains whether non-perturbative type effects can generate the necessary clockwork structure in type IIB. Heuristically, we expect non-perturbative effects to enter the theory through the superpotential as

$$W_{np} = \sum_i A_i e^{a_i^j T_j} \,, \tag{10.1}$$

where i runs over the number of E3-branes in the theory, then by appropriately choosing a basis a_i^j in instanton parameter space, one would hope to generate the right structure if the internal geometry is anisotropic, i.e. there are some small and large cycle such that a hierarchy $\tau_b \gg \tau_i$ for all $i \neq b$ exists. This would allow to single out a Kähler direction with an associated axion much lighter than the other directions. Although this picture is attractive, obtaining a full ten-dimensional description is not trivial. For example writing down the required basis a_i^j could, a priori, lead to inconsistencies on the ten-dimensional theory. An example of such inconsistencies is given by strong forms of the weak gravity conjecture [1, 2].

In Chap. 5 we concentrated on the coincidence problem and its implications for dynamical models of dark energy obtained from generic properties of string compactifications close to the boundary of moduli space, where we assumed that the dynamics for the quintessence field are described by some runaway potential toward a Dine-Seiberg type vacuum at large vevs. We were able to show that, due to the mass of the heavy states depending on the runaway quintessence field, particle production is enhanced over time. This allows us to set an upper bound on the time it takes the particle production to swamp the effective description. We found that in some $\mathcal{O}(100)H_0$ time, the particle production becomes too much and the quintessence regime is destabilised. This ameliorated the coincidence problem to a 1-in-100 chance to find ourselves close to the dawn of the dark energy domination epoch. Finally, we argued that a more stringent bound might be obtained from a true Calabi-Yau compactification of the supergravity action, rather than a (well) motivated EFT.

Before moving on to the second half of the thesis, we spent some time in Chap. 6 discussing the differences between supergravity computations at the boundary of moduli space and at the bulk of moduli space. We argued that, as long as the couplings are small enough (but finite) in the bulk of moduli space[2] the perturbative expansion is well-posed and this allows us to work in a region of moduli space that is phenomenologically interesting. Furthermore, we were able to prove that on the run-away of moduli space one cannot achieve a slow-roll regime. This, together with the usual no-go of [3], prevents us from obtaining a phenomenologically viable description of our Universe. We pointed out that this implies that there exists a disconnect from the phenomenological claims of the Swampland programme and cosmology.

[2] This is to say that we are at large volume and weak coupling, rather than deep in the bulk of moduli space where the volume can be small or the physics being strongly coupled.

Moving on to the second half of the thesis, in Chap. 7 we introduced a number of perturbative and non-perturbative corrections that are typically used to model build de Sitter vacua in type IIB supergravity. We reviewed each of these in some detail, as well as the best understood effective de Sitter constructions in type IIB, namely KKLT, the racetrack models and the large volume scenario. Armed with this knowledge, we tried to construct a viable model of quintessence in Chap. 8. However, we found when considering the cosmological evolution of our Universe, from an early period of inflation to late time dark energy, that a mixture between the Kallosh-Linde problem and quantum diffusion allowed only axionic hilltop models of quintessence with an extreme fine of the initial conditions. We argued that these constraints make quintessence less attractive to model builders than a cosmological constant. Therefore, without a clear input from cosmological data, i.e. finding with some degree of confidence an equation of state parameter for dark energy different than unity, string phenomenologists are better off focusing on de Sitter vacua.

Moving forward, there are two clear directions. One is to continue advancing our understanding of the de Sitter constructions that have already been developed by the community. This will require a better handle on perturbative and non-perturbative corrections, as well as, a clear ten-dimensional picture of all the necessary ingredients to generate such solution. A second option would be to introduce new ingredients that generate new solutions, like non-linearly realised supersymmetry which is the main idea behind the de Sitter supergravity constructions of [4, 5]. These new solutions might be the key to a more complete picture of supergravity.

In Chap. 9 we follow the former approach. We consider runaway effective field theory that have some connection with fluxed supergravity and show the effects of higher curvature corrections to the leading order solution. In particular, we considered the Gauss-Bonnet corrections as motivated in the heterotic supergravity framework. We found that, on the boundary of parameter space where the leading order theory would yield an external four-dimensional space, the higher order corrections contributed with negative curvature thus generating anti-de Sitter solutions. Chapter 9 served to point out the power of the effective theory approach with regards to calculating higher order corrections. This approach can be very helpful in flagging interesting model building terms. However, some notion of the form of the corrections is known. Of course, historically it has been very hard to obtain any specifics about the behaviour of these corrections in any generality. A recent and commendable effort in this direction is [6], where the existence and volume scaling of α'-corrections is explored.

Another interesting and qualitatively different approach is the one followed by [7, 8]. In this case, the authors take advantage of the AdS/CFT correspondence to study the holographic behaviour of the AdS vacua of a number of supergravity models, where the Kähler moduli can be stabilised consistently (like KKLT or LVS). Interestingly, they find that each of these models presents a certain universality in the conformal dimensions of the CFT operators dual to the moduli fields in the AdS vacua. This leads the authors to speculate about the origin of this universality. If true for any AdS construction, it would seem to imply that the a priori vast string landscape maps to a limited set of conformal dimensions when translated into the CFT language, which are, for example, independent of choices of fluxes.

In this thesis, we have concerned ourselves with the study of the microscopic origin of our Universe with a focus on the nature of dark energy. In the first half of the thesis, we considered the consequence of some stringy motivated effective field theory on naturalness and the coincidence problem. In the second half, we got our hands dirty working with the effective field theory descending from type IIB supergravity. Although we were able to provide some very strong phenomenological results, sadly, we do not find ourselves any closer to a microscopic description of our Universe. We are able to say what doesn't work but lack an intuition of what might work from a string theory perspective. However, one should not be despondent. This might just be because a more complete picture of the non-perturbative aspects of string theory is needed, or maybe it requires new interpretations of supergravity.

The microscopic origin of our Universe is a very challenging open problem which has provided a concrete framework to test and better understand string theory. To some extent, this could be seen as one of the last great challenges of theoretical physics. In a somewhat related position we would have found Michelson and Kelvin at the turn of the 20th century.[3] In [9], Michelson claimed the following,

> The more important fundamental laws and facts of physical science have all been discovered, and these are now so firmly established that the possibility of their [sic.] ever being supplanted in consequence of new discoveries is exceedingly remote.

which is, in hindsight, a *brave* claim. In [10], Kelvin would go on to point out the "two clouds over the dynamical theory of heat and light". These two clouds where related to the nature of ether and the partition energy as set out by Maxwell and Boltzmann and the specific heat of gases. The first cloud would be dispersed by special relativity and the second one by quantum physics, in the meantime generating storms of which this thesis is just a small part. It is worth noting that in [9], Michelson would go on to add the now famous quote "our future discoveries must be looked for in the sixth place of decimals", which would turn out to be true, albeit a bit misdirected when taking into consideration his former quote on the laws of physics.

As some concluding words to this thesis, let us see what appears after the clouds covering the cosmological constant are dispersed and which fundamental laws need rewriting. It is certainly a very exciting time for Physics.

References

1. Brown J, Cottrell W, Shiu G, Soler P (2016) On axionic field ranges, loopholes and the weak gravity conjecture. JHEP 04:017. https://doi.org/10.1007/JHEP04(2016)017, arxiv:https://arxiv.org/abs/1504.00659
2. Shiu G, Soler P, Cottrell W (2019) Weak gravity conjecture and extremal black holes. Sci ChinaPhys Mech Astron 62:110412. https://doi.org/10.1007/s11433-019-9406-2, arxiv:https://arxiv.org/abs/1611.06270

[3] I would like to thank Prof. Javier Tejada for bringing this anecdote to my attention some years ago while I was an undergraduate.

3. Maldacena JM, Nunez C (2001) Supergravity description of field theories on curved manifolds and a no go theorem. Int J Mod Phys A 16:822. https://doi.org/10.1142/S0217751X01003937, arxiv:https://arxiv.org/abs/hep-th/0007018hep-th/0007018
4. Bergshoeff EA, Freedman DZ, Kallosh R, Van Proeyen A (2015) Pure de Sitter supergravity. Phys Rev D 92:085040. https://doi.org/10.1103/PhysRevD.93.069901, arxiv:https://arxiv.org/abs/1507.08264
5. Bergshoeff E, Freedman D, Kallosh R, Van Proeyen A (2016) Construction of the de Sitter supergravity. In: About various kinds of interactions: workshop in honour of Professor Philippe Spindel, 2. arxiv:https://arxiv.org/abs/1602.01678
6. Cicoli M, Quevedo F, Savelli R, Schachner A, Valandro R (2021) Systematics of the α' expansion in F-theory. JHEP 8:99. https://doi.org/10.1007/JHEP08(2021)099, arxiv:https://arxiv.org/abs/2106.04592
7. Conlon JP, Revello F (2020) Moduli stabilisation and the holographic swampland. LHEP 171. https://doi.org/10.31526/lhep.2020.171, arxiv:https://arxiv.org/abs/2006.01021
8. Conlon JP, Ning S, Revello F, Exploring the holographic swampland. https://arxiv.org/abs/2110.062452110.06245
9. Michelson A (1903) Light waves and their uses. Chicago University. The Decennial Publications. 2d series, University of Chicago Press
10. Kelvin RHL (1901) I. nineteenth century clouds over the dynamical theory of heat and light. Lond Edinb Dublin Philos Mag J Sci 2:1. https://doi.org/10.1080/14786440109462664, arxiv:https://arxiv.org/abs/https://doi.org/10.1080/14786440109462664

Appendix A
Details and Calculations for Type IIB Supergravity

In this appendix, we write the details of some of the least enlightening and more annoying calculations for Chap. 2. We will describe the details of the dimensional reduction of the $\mathcal{N} = 2$ type IIB action and discuss the dualisation of the Kalb-Ramond and RR two-form axion.

A.1 Calculations for the Dimensional Reduction of the $\mathcal{N} = 2$ Type IIB Action

We would like to find the dimensional reduction of Eq. (2.39). It is most convenient to work in string frame and do an overall Weyl rescaling at the end. Therefore, our starting point is the string frame action

$$\begin{aligned} S_{IIB} = &\int \frac{e^{-2\hat{\phi}}}{2}\left(\hat{R} \star \mathbf{1} + 4d\hat{\phi} \wedge \star d\hat{\phi} - \hat{H}_3 \wedge \star \hat{H}_3\right) \\ &- \int \frac{1}{4}\left(\hat{F}_1 \wedge \star \hat{F}_1 + \hat{F}_3 \wedge \star \hat{F}_3 + \frac{1}{2}\hat{F}_5 \wedge \star \hat{F}_5 + \hat{C}_4 \wedge \hat{H}_3 \wedge \hat{F}_3\right) \end{aligned} \tag{A.1}$$

where the field strength are Eq. (2.40)

$$\begin{aligned} \hat{H}_3 = d\hat{B}_2, \quad \hat{F}_1 = d\hat{C}_0, \quad \hat{F}_3 = d\hat{C}_2 + \hat{B}_2 \wedge d\hat{C}_0, \\ \hat{F}_5 = d\hat{C}_4 + \hat{B}_2 \wedge d\hat{C}_2, \end{aligned} \tag{A.2}$$

and with the field content given in Eq. (2.41)

$$\begin{aligned} \hat{B}_2 = B_2(x) + b^A(x)\omega_A, \quad \hat{C}_2 = C_2(x) + c_2^A\omega_A, \\ \hat{C}_4 = c_4^A(x)\omega_A + \mathcal{X}i^{\hat{K}}(x) \wedge \alpha_{\hat{K}} - \tilde{\xi}_{\hat{K}}(x) \wedge \beta^{\hat{K}} + \rho_A(x)\tilde{\omega}^A. \end{aligned} \tag{A.3}$$

F. Cunillera, *Dark Energy*, Springer Theses,
https://doi.org/10.1007/978-3-031-21468-4

Appendix B
Type IIA Perturbative Supergravity

In this appendix we provide details on how to obtain the $\mathcal{N} = 1$ scalar potential for fluxed type IIA supergravity. We will begin by finding the analogue of the type IIB $\mathcal{N} = 2$ scalar potential. Then, we will see how the orientifold projection works for type IIA theories and we will present the orientifolded scalar potential. Lastly, we will discuss moduli stabilisation in type IIA theories.

B.1 $\mathcal{N} = 2$ Supergravity Action

In spirit, this section follows very closely the analogue type IIB calculation in Sect. 2.2. Again, it is convenient to work in the string frame and do an overall Weyl rescaling at the end. We begin by writing down the 10-dimensional type IIA supergravity action in string frame [1]

$$\begin{aligned} S_{IIA} = & \int \frac{e^{-2\hat{\phi}}}{2} \left(\hat{R} \star \mathbf{1} + 4 d\hat{\phi} \wedge \star d\hat{\phi} - \frac{1}{2} \hat{H}_3 \wedge \star \hat{H}_3 \right) \\ & - \frac{1}{2} \int \left(\hat{F}_2 \wedge \star \hat{F}_2 + \hat{F}_4 \wedge \star \hat{F}_4 + \mathcal{L}_{\text{top}} \right), \end{aligned} \tag{B.1}$$

where the NSNS sector is the same as in the type IIB theory with the ten-dimensional dilaton $\hat{\phi}$ and the Kalb-Ramond two-form $\hat{B}_2$, whereas the RR sector now contains odd-form gauge fields whose field strengths we define as follows

$$\hat{F}_2 = d\hat{C}_1, \quad \hat{F}_4 = d\hat{C}_3 - d\hat{C}_1 \wedge \hat{B}_2. \tag{B.2}$$

where $(\hat{B}_2)^p := \underbrace{\hat{B}_2 \wedge \cdots \wedge \hat{B}_2}_{p-\text{times}}$. The topological term is given by

F. Cunillera, *Dark Energy*, Springer Theses,
https://doi.org/10.1007/978-3-031-21468-4

Table B.1 Bases for a generic Calabi-Yau threefold

Cohomology group	Dimension	Basis
$H^{1,1}$	$h^{1,1}$	ω_A
$H^{2,2}$	$h^{1,1}$	$\tilde{\omega}^A$
H^3	$2+2h^{2,1}$	$\left(\alpha_{\hat{K}}, \beta^{\hat{L}}\right)$
$H^{2,1}$	$h^{2,1}$	χ_K
$H^{3,3}$	1	$\mathrm{dvol}(\mathcal{X})$

$$\mathcal{L}_{\text{top}} = \hat{B}_2 \wedge d\hat{C}_3 \wedge d\hat{C}_3 - (\hat{B}_2)^2 \wedge d\hat{C}_3 \wedge d\hat{C}_1 + \frac{1}{3}(\hat{B}_2)^3 \wedge d\hat{C}_1 \wedge d\hat{C}_1. \tag{B.3}$$

We can expand in the bases of harmonic (p,q)-forms that we found in Sect. 2.1 and that we summarise below in Table B.1 for convenience. The fields are given in these bases by

$$\hat{B}_2 = B_2(x) + b^A\omega_A, \quad \hat{C}_1 = A^0(x), \quad \hat{C}_3 = C_3(x) + A^A(x) \wedge \omega_A + \xi^{\hat{L}}\alpha_{\hat{L}} + \tilde{\xi}_{\hat{K}}\beta^{\hat{K}}. \tag{B.4}$$

Here, we have used notation that is reminiscent of the type IIB description, since the moduli will fill similar roles to their IIB counterparts. Let us analyse the dimensional reduction of the type IIA action Eq. (B.1). As in the type IIB case, we will take the ansatz for the ten-dimensional metric to be

$$ds^2 = g_{\mu\nu}(x)dx^\mu dx^\nu + g_{i\bar{j}}(x, z)dz^i d\bar{z}^{\bar{j}}, \tag{B.5}$$

where $g_{\mu\nu}$ (with $\mu, \nu = 0, \ldots, 3$) is the external metric and $g_{i\bar{j}}$ (with $i, \bar{j} = 1, \ldots, 3$) is the CY metric of $\mathcal{X}$.

The reduction of the NSNS sector is identical to the type IIB case, and we refer the reader to Sect. A.1.1.

B.1.1 The RR Sector and the Topological Term of the Type IIA Action

Below, we plug the field content Eq. (B.4) into the different terms of the R-R sector. For the 2-form we find

$$-\frac{1}{2}\int_{\mathcal{X}} \hat{F}_2 \wedge \star\hat{F}_2 = -\frac{\mathcal{V}}{2} dA^0 \wedge \star_4 dA^0, \tag{B.6}$$

where $\mathcal{V}$ is the volume of $\mathcal{X}$ and G_{AB} is given by Eq. (2.20). The 4-form dimensional reduction reads

$$-\frac{1}{2}\int_{\mathcal{X}} \hat{F}_4 \wedge \star\hat{F}_4 = -\frac{1}{2}\left[\mathcal{V}\left(dC_3 - B_2 \wedge dA^0\right) \wedge \star_4 \left(dC_3 - B_2 \wedge dA^0\right)\right]$$
$$+ 4\mathcal{V}G_{AB}\left(dA^A - b^A dA^0\right) \wedge \star_4 \left(dA^B - b^B dA^0\right)$$
$$- (\mathrm{Im}\mathcal{M})^{-1\,\hat{K}\hat{L}}\left(d\tilde{\xi}_{\hat{K}} + \mathcal{M}_{\hat{K}\hat{M}}d\xi^{\hat{M}}\right) \wedge \star_4 \left(d\tilde{\xi}_{\hat{L}} + \bar{\mathcal{M}}_{\hat{L}\hat{N}}d\xi^{\hat{N}}\right), \tag{B.7}$$

where we have used $G^{AB} := 4\mathcal{V}\int \tilde{\omega}^A \wedge \star\tilde{\omega}^B$ is the inverse metric of G_{AB}, as well as the period matrix defined in Eq. (2.35) and given below for convenience

$$\int \alpha_{\hat{K}} \wedge \star\alpha_{\hat{L}} = -\left[\mathrm{Im}\,\mathcal{M} + (\mathrm{Re}\,\mathcal{M})(\mathrm{Im}\,\mathcal{M})^{-1}(\mathrm{Re}\,\mathcal{M})\right]_{\hat{K}\hat{L}}, \tag{B.8}$$

$$\int \beta^{\hat{K}} \wedge \star\beta^{\hat{L}} = -(\mathrm{Im}\,\mathcal{M})^{-1\,\hat{K}\hat{L}}, \tag{B.9}$$

$$\int \alpha_{\hat{K}} \wedge \star\beta^{\hat{L}} = -\left[(\mathrm{Re}\,\mathcal{M})(\mathrm{Im}\,\mathcal{M})^{-1}\right]_{\hat{K}}^{\hat{L}}. \tag{B.10}$$

Finally the topological term is

$$\int_{\mathcal{X}} \hat{B}_2 \wedge d\hat{C}_3 \wedge d\hat{C}_3 - (\hat{B}_2)^2 \wedge d\hat{C}_3 \wedge d\hat{C}_1 + \frac{1}{3}(\hat{B}_2)^3 \wedge d\hat{C}_1 \wedge d\hat{C}_1$$
$$= B_2 \wedge d\left(\tilde{\xi}_{\hat{K}}d\xi^{\hat{K}} - \xi^{\hat{L}}d\tilde{\xi}_{\hat{L}}\right) + \mathcal{K}_{ABC}b^A\left[dA^B \wedge dA^C - b^B dA^0 \wedge \left(dA^C - \frac{1}{3}b^C dA^0\right)\right] \tag{B.11}$$

The action can now be written in terms of the gauge-coupling matrices defined in Eq. (2.24)

$$\mathrm{Re}\mathcal{N} = \begin{pmatrix}-\frac{1}{3}\mathcal{K}_{ABC}b^Ab^Bb^C & \frac{1}{2}\mathcal{K}_{ABC}b^Ab^B \\ \frac{1}{2}\mathcal{K}_{ABC}b^Ab^B & -\mathcal{K}_{ABC}b^C\end{pmatrix}, \tag{B.12}$$

$$\mathrm{Im}\mathcal{N} = -\mathcal{V}\begin{pmatrix}1 + 4G_{AB}b^Ab^B & -4G_{AB}b^B \\ -4G_{AB}b^B & 4G_{AB}\end{pmatrix}, \tag{B.13}$$

$$(\mathrm{Im}\mathcal{N})^{-1} = -\frac{1}{\mathcal{V}}\begin{pmatrix}1 & b^A \\ b^A & \frac{1}{4}G^{AB} + b^Ab^B\end{pmatrix}. \tag{B.14}$$

Before doing so, we will first dualise the B_2 and C_3 fields into a scalar $a(x)$ and a constant flux e_0, respectively. This will allow us to put the $\mathcal{N} = 2$ action in its gauged form [2], turning a double tensor multiplet containing B_2 into a hypermultiplet in terms of a.

B.1.2 Dualising the B_2 and C_3 Fields

Let us write the Lagrangians we are trying to dualise explicitly and work from there. Starting from the C_3 gauge field, we collect all pieces containing C_3 in

$$\begin{aligned}\mathcal{L}_{C_3} &= -\frac{\mathcal{V}}{2}\left(dC_3 - B_2 \wedge dA^0\right) \wedge \star_4 \left(dC_3 - B_2 \wedge dA^0\right) \\ &:= -\frac{\mathcal{V}}{2}\left(dC_3 - J_4\right) \wedge \star_4 \left(dC_3 - J_4\right) - e_0 \left(dC_3 - 0\right),\end{aligned} \tag{B.15}$$

where we have introduced J_4 for notational convenience. The Lagrange multiplier e_0 enforces the (trivial) Bianchi identity.[3] The equation of motion for dC_3 becomes

$$\frac{\mathcal{V}}{2} \star_4 \left(dC_3 - J_4\right) = -e_0 \quad \Rightarrow \quad dC_3 = J_4 + \frac{e_0}{\mathcal{V}} \star_4 \mathbf{1} \tag{B.16}$$

and substituting back

$$\mathcal{L}_{e_0} = -\frac{e_0^2}{2\mathcal{V}} \star_4 \mathbf{1} - e_0 J_4 = -\frac{e_0^2}{2\mathcal{V}} \star_4 \mathbf{1} - e_0 \left(B_2 \wedge dA^0\right). \tag{B.17}$$

Next, we can substitute the expression back into the Lagrangian $\mathcal{L}_{C_3}$ to obtain the dual description. We also choose to group the field strength fields present in the theory as

$$F^{\hat{A}} := \{dA^0, dA^A\}, \tag{B.18}$$

for later convenience. Finally, we would like to dualise the dB_2 field strength into the scalar $a(x)$. This follows from our results in Sect. A.2, which we recall below for convenience. In this case, the Lagrangian for the H_3 field strength reads

$$\mathcal{L}_{H_3} = -\frac{1}{4} dB_2 \wedge \star dB_2 + \frac{1}{2} dB_2 \wedge \left(\tilde{\xi}_{\hat{K}} d\xi^{\hat{K}} - \xi^{\hat{L}} d\tilde{\xi}_{\hat{L}}\right), \tag{B.19}$$

the equation of motion for B_2 implies

$$d\left[\frac{1}{2} \star H_3 - \left(\tilde{\xi}_{\hat{K}} d\xi^{\hat{K}} - \xi^{\hat{L}} d\tilde{\xi}_{\hat{L}}\right)\right] = 0 \quad \Rightarrow \quad da = \frac{1}{2} \star dB_2 - \left(\tilde{\xi}_{\hat{K}} d\xi^{\hat{K}} - \xi^{\hat{L}} d\tilde{\xi}_{\hat{L}}\right), \tag{B.20}$$

which, together with Eqs. (B.6), (B.7), (B.11), (B.16), can be substituted in Eq. (B.1) to find the four-dimensional type IIA effective action[4] [2]

---

[3] Note that if one tries to simply find the equation of motion for $\star C_3$ and substitute back, because dC_3 is a top-form then the equations of motion would contain a derivative of a top-form which must vanish identically and this method does not work. One must introduce the Lagrange multiplier in the case of a top-form field strength.

Table B.2 Multiplets for the 4-dimensional type IIB supergravity spectrum

Multiplet	Dimension	Field content
Gravity multiplet	1	$(g_{\mu\nu}, A^0)$
Vector multiplet	$h^{1,1}$	(A^A, v^A, b^A)
Hypermultiplet	$h^{2,1}+1$	$\left(\xi^K, \tilde{\xi}_K, z^K\right) + \left(a, \phi, \xi^0, \tilde{\xi}_0\right)$

$$S_{IIA}^{(4)} = \int -\frac{1}{2}R \star \mathbf{1} + \frac{1}{2}\mathrm{Im}\mathcal{N}_{\hat{A}\hat{B}}F^{\hat{A}} \wedge \star F^{\hat{B}} + \frac{1}{2}\mathrm{Re}\mathcal{N}_{\hat{A}\hat{B}}F^{\hat{A}} \wedge F^{\hat{B}}$$
$$- G_{AB}dT^A \wedge \star d\bar{T}^B - h_{uv}d\tilde{q}^u \wedge \star d\tilde{q}^v - V \star \mathbf{1}, \tag{B.21}$$

with

$$V = \frac{e^{4\phi}}{2\mathcal{V}}e_0^2, \tag{B.22}$$

and

$$h_{uv}d\tilde{q}^u \wedge \star d\tilde{q}^v := d\phi \wedge \star d\phi + G_{K\bar{L}}dz^K \wedge \star d\bar{z}^L +$$
$$\frac{e^{4\phi}}{4}\left(Da + \tilde{\xi}_{\hat{K}}d\xi^{\hat{K}} - \xi^{\hat{L}}d\tilde{\xi}_{\hat{L}}\right) \wedge \star \left(Da + \tilde{\xi}_{\hat{K}}d\xi^{\hat{K}} - \xi^{\hat{L}}d\tilde{\xi}_{\hat{L}}\right)$$
$$-\frac{e^{2\phi}}{2}(\mathrm{Im}\mathcal{M})^{-1\,\hat{K}\hat{L}}\left(d\tilde{\xi}_{\hat{K}} + \mathcal{M}_{\hat{K}\hat{M}}d\xi^{\hat{M}}\right) \wedge \star \left(d\tilde{\xi}_{\hat{L}} + \bar{\mathcal{M}}_{\hat{L}\hat{N}}d\xi^{\hat{N}}\right), \tag{B.23}$$

where $Da := da + 2e_0A^0$. We have dropped the subindex on the four-dimensional Hodge star, since there are no more ambiguities. Notably, the scalar potential of the type IIA $\mathcal{N}=2$ is not flat due to the presence of the Freund-Rubin *flux* e_0, which is the gauge charge for the axion a dual to the external piece of the Kalb-Ramond two-form. The spectrum of the theory is given in Table B.2.

B.2 Orientifolding and $\mathcal{N}=1$ Type IIA Supergravity

Similar to the type IIB case, we must reduce the supergravity spectrum by halving the number of supersymmetry generators through an orientifold projection.

We write down the orientifold projection operator as [3]

$$\mathcal{O} = \Omega_p\,(-1)^{\mathrm{F_L}}\,\sigma. \tag{B.24}$$

[4] Where a similar Weyl rescaling has been done to put the action in Einstein frame, similar to the one done on the IIB side.

Table B.3 Harmonic bases for the orientifolded $\mathcal{X}$

Cohomology group	Dimension	Basis
$H^{1,1}_{+}$	$h^{1,1}_{+}$	ω_α
$H^{1,1}_{-}$	$h^{1,1}_{-}$	ω_a
$H^{2,2}_{+}$	$h^{1,1}_{-}$	$\tilde{\omega}^a$
$H^{2,2}_{-}$	$h^{1,1}_{+}$	$\tilde{\omega}^\alpha$
H^3_{+}	$h^{2,1}+1$	$(\alpha_{\hat{k}}, \beta^\lambda)$
H^3_{-}	$h^{2,1}+1$	$\left(\alpha_\lambda, \beta^{\hat{k}}\right)$

If one wishes to preserve $\mathcal{N}=1$ supersymmetry, then the involution must act on the Kähler form and the holomorphic 3-form as

$$\sigma^* J = -J, \qquad \sigma^*\Omega = e^{2i\theta}\bar{\Omega}, \tag{B.25}$$

for a constant phase $e^{2i\theta}$ and where σ^* is the pullback of σ, an anti-holomorphic involution acting on $\mathcal{X}$. This in contrast with the type IIB orientifold conditions where the involution is holomorphic. In type IIA supergravity, the above conditions are consistent with the introduction of $O6$-planes in the theory.

We have seen that the effect of the involution on the cohomology of $\mathcal{X}$ is to split them in even and odd cohomologies

$$H^{p,q} = H^{p,q}_{+} \oplus H^{p,q}_{-}, \tag{B.26}$$

such that an element $\omega_\pm \in H^{p,q}_{\pm}$ transforms like $\sigma^*\omega_\pm = \pm\omega_\pm$. From the type IIA orientifold conditions Eq. (B.25), we can further see that

- $\sigma^* J = -J$ implies $h^{3,3}_{-} = 1$, since $\mathcal{V} \propto \int J \wedge J \wedge J$ and $h^{0,0}_{+} = 1$, by Hodge duality. Alternatively, this implies that $h^{3,3}_{+} = h^{0,0}_{-} = 0$.
- Due to the anti-holomorphicity of σ and Hodge duality, we find $h^{1,1}_{\pm} = h^{2,2}_{M_{\text{pl}}}$ and $h^3_{+} = h^3_{-} = h^{2,1}+1$.

In Table B.3 we summarise the notation for the truncated harmonic bases after orientifolding, which largely follows the notation of the type IIB analogues,

where $\hat{k} = 0, \ldots, \tilde{h}$ and $\lambda = \tilde{h}, \ldots, h^{2,1}$. The non-trivial intersection numbers are given by

$$\int_{\mathcal{X}} \omega_a \wedge \tilde{\omega}^b = \delta^b_a, \qquad \int_{\mathcal{X}} \omega_\alpha \wedge \tilde{\omega}^\beta = \delta^\beta_\alpha, \qquad \int_{\mathcal{X}} \alpha_{\hat{k}} \wedge \beta^{\hat{l}} = \delta^{\hat{l}}_{\hat{k}}, \qquad \int_{\mathcal{X}} \alpha_\kappa \wedge \beta^\lambda = \delta^\lambda_\kappa. \tag{B.27}$$

The action of the parity operator and the fermion number operator on the 10-dimensional fields is summarised in Table B.4 [4]. It is clear from it that the invariant orientifolded states must obey

Table B.4 Actions of the parity and fermion number operators on the 10-dimensional spectrum.

Field	Under Ω_p	Under $(-1)^{F_L}$	Under $\Omega_p\,(-1)^{F_L}$
$\hat{\phi}$	+	+	+
$\hat{g}$	+	+	+
$\hat{B}_2$	−	+	−
$\hat{C}_1$	+	−	−
$\hat{C}_3$	−	−	+

$$\sigma^*\hat{\phi} = +\hat{\phi}, \quad \sigma^*\hat{g} = +\hat{g}, \quad \sigma^*\hat{B}_2 = -\hat{B}_2, \\ \sigma^*\hat{C}_1 = -\hat{C}_1, \quad \sigma^*\hat{C}_3 = +\hat{C}_3. \tag{B.28}$$

One can truncate the $\mathcal{N} = 2$ spectrum as follows. For the Kähler structure we obtain

$$J = v^a\omega_a, \quad B_2 = b^a\omega_a, \quad J_c := J - iB_2 = T^a\omega_a := (v - ib)^a\omega_a. \tag{B.29}$$

Note, in particular, that the orientifold has projected out the external $B_2(x)$ piece, which before we had dualised into a gauge scalar charged under e_0. The complex structure deformations are given by applying the orientifold constraint to obtain the conditions

$$\mathrm{Im}(C\mathcal{Z}^{\hat{k}}) = \mathrm{Re}(C\mathcal{F}_{\hat{k}}) = 0, \quad \mathrm{Re}(C\mathcal{Z}^{\lambda}) = \mathrm{Im}\,(C\mathcal{F}_{\lambda}) = 0 \qquad C := e^{-\phi-i\theta}e^{K^{cs}/2} \tag{B.30}$$

so that the expansion of the holomorphic three-form with the regulator function C is

$$C\Omega = \mathrm{Re}(C\mathcal{Z}^{\hat{k}})\alpha_{\hat{k}} + i\mathrm{Im}(C\mathcal{Z}^{\lambda})\alpha_{\lambda} - \mathrm{Re}(C\mathcal{F}_{\lambda})\beta^{\lambda} - i\mathrm{Im}(C\mathcal{F}_{\hat{k}})\beta^{\hat{k}}. \tag{B.31}$$

One can still choose the Kähler gauge so that $\mathcal{Z}^0 = 1$, however it is most convenient to keep the gauge freedom for now while keeping in mind that there are only $h^{2,1}$ degrees of freedom in the complex structure sector. Finally, let us look at the one and three-form gauge field in the RR sector.

The involution constraints impose that

$$\sigma^*\hat{A}_1 = -\hat{A}_1, \quad \rightarrow \quad \hat{A}_1 \in H^{1,0}(Y). \tag{B.32}$$

If $\mathcal{X}$ is a Calabi-Yau threefold, there are no non-trivial one-cycles and $\hat{A}_1$ should at most be an external field. However, the orientifold acts trivially on any external fields, thus the $\mathcal{N} = 2$ graviphoton A_0 must be projected out.

Finally,

$$\sigma^*\hat{C}_3 = \hat{C}_3, \quad \rightarrow \quad \hat{C}_3 \in H^3_+(Y). \tag{B.33}$$

In this case, $\hat{C}_3$ admits an external form C_3, that will dualise once again into the Freud-Rubin flux e_0, such that the gauge field can be expanded into

$$\hat{C}_3 = C_3(x) + A^\alpha \wedge \omega_\alpha + \xi^{\hat{k}} \alpha_{\hat{k}} - \tilde{\xi}_\kappa \beta^\kappa := C_3(x) + A^\alpha \wedge \omega_\alpha + C_3^i(x) \quad \text{(B.34)}$$

defining a complexified three-form

$$\Omega_c := C_3^i + 2i\text{Re}\,(C\Omega)\,, \qquad C := e^{-\phi - i\theta} e^{K^{cs}/2}. \quad \text{(B.35)}$$

Including the dilaton through the compensator function C is equivalent to trading the irrelevant scale factor of the three-form Ω by the dilaton, so that all $h^{2,1} + 1$ fields in the definition of Ω_c are physical. This will make it so introducing coordinates to the quaternionic manifold later on can be done directly through Ω_c.

Much like in type IIB analogue, locally the moduli space can still be written in a product form of two Kähler manifolds

$$\tilde{\mathcal{M}}^{ks} \times \tilde{\mathcal{M}}^{Q}, \quad \text{(B.36)}$$

with $\tilde{\mathcal{M}}^{ks}$ and $\tilde{\mathcal{M}}^{Q}$ subspaces of the $\mathcal{N} = 2$ Kähler and quaternionic manifolds, respectively. In contrast with the type IIB, the complex structure manifold is now part of the quaternionic manifold, $\mathcal{M}^{cs} \in \mathcal{M}^{Q}$. This would be expected from the usual duality arguments, where the Kähler sector of type IIB maps to a dual complex structure sector in IIA and vice versa (for explicit details of the mirror symmetry see, for example [3, 5], and references therein). Now we would like to study the effects of the orientifold projection onto the Kähler and quaternionic sector following the conditions laid out before to obtain the $\mathcal{N} = 1$ effective action.

B.2.1 The Kähler Structure of $\tilde{\mathcal{M}}^{ks}$

The number of Kähler moduli $h^{1,1}$ is reduced in half to $h_-^{1,1}$, *i.e.*

$$t^A \to t^a, \qquad \omega_A \to \omega_a, \qquad a = 1, \ldots, h_-^{1,1}. \quad \text{(B.37)}$$

Note that these projections imply some vanishing of the triple intersection numbers. Indeed, recall that $d\mathcal{V} \in H_-^6$ which means that

$$\mathcal{K}_{ABC} v^A v^B v^C := \int_{\mathcal{X}} v^A v^B v^C \omega_A \wedge \omega_B \wedge \omega_C \propto \int_{\mathcal{X}} d\mathcal{V}, \quad \text{(B.38)}$$

so the integrand must be odd under the action of σ and thus only intersection numbers with one or three odd-legs are non-trivial after the projection. Similarly, one can apply the same reasoning to the *contractions* of the intersection numbers, *i.e.* $\mathcal{K}_{AB}$ and $\mathcal{K}_A$, together with the fact that $J \in H_-^{1,1}$ to find

$$\mathcal{K}_{\alpha\beta\gamma} = \mathcal{K}_{\alpha ab} = \mathcal{K}_{\alpha a} = \mathcal{K}_{\alpha} = 0. \tag{B.39}$$

Then, the $\mathcal{N} = 1$ Kähler metric can be obtained directly by using the conditions above on the $\mathcal{N} = 2$ Kähler metric in Eq. (2.20). The orientifolded metric takes the block-diagonal form

$$G_{ab} = -\frac{1}{4\mathcal{V}}\left(\mathcal{K}_{ab} - \frac{\mathcal{K}_a\mathcal{K}_b}{4\mathcal{V}}\right), \qquad G_{\alpha\beta} = -\frac{\mathcal{K}_{\alpha\beta}}{4\mathcal{V}}, \tag{B.40}$$

$$G_{\alpha a} = 0. \tag{B.41}$$

The same constraints can be applied to the gauge-kinetic coupling matrix $\mathcal{N}_{\hat{A}\hat{B}}$ to get

$$\mathrm{Re}\mathcal{N}_{\alpha\beta} = -\mathcal{K}_{\alpha\beta a}b^a, \qquad \mathrm{Im}\mathcal{N}_{\alpha\beta} = \mathcal{K}_{\alpha\beta}, \tag{B.42}$$

$$\mathcal{N}_{0\alpha} = \mathcal{N}_{a\alpha} = 0, \tag{B.43}$$

plus the components $\mathcal{N}_{\hat{a}\hat{b}}$ found by substituting $\left(\hat{A}, \hat{B}, \hat{C}\right) \to \left(\hat{a}, \hat{b}, \hat{c}\right)$, with $\hat{a} = 0, a$, in the formulae of Eqs. (2.25)–(2.27). Finally, the Kähler potential is simply given by

$$K^{ks} = -\ln\left[\frac{1}{6}\mathcal{K}_{abc}\left(T + \bar{T}\right)^a\left(T + \bar{T}\right)^b\left(T + \bar{T}\right)^c\right] = -\ln\left(\bar{T}^{\hat{a}}\partial_{T^{\hat{a}}}F - T^{\hat{a}}\partial_{\bar{T}^{\hat{a}}}\bar{F}\right), \tag{B.44}$$

with $\hat{a} = 0, 1, \ldots, h_-^{1,1}$ the adapted coordinates of $\mathcal{M}^{ks}$ and the prepotential function that generates the Kähler potential is $F = \frac{1}{3}\mathcal{K}_{abc}\frac{T^aT^bT^c}{T^0}$. The $\mathcal{N} = 1$ Kähler sector for the orientifolded effective action reads

$$S_{IIA}^{(4)} = \int -G_{ab}\, dT^a \wedge \star d\bar{T}^b + \frac{1}{2}\mathrm{Im}\mathcal{N}_{\alpha\beta}F^\alpha \wedge \star F^\beta + \frac{1}{2}\mathrm{Re}\mathcal{N}_{\alpha\beta}F^\alpha \wedge F^\beta, \tag{B.45}$$

with $F^\alpha := dA^\alpha$.

B.2.2 The Quaternionic Structure of $\tilde{\mathcal{M}}^Q$

The quaternionic sector is more involved. The complex structure deformations complying with the orientifolds constraints are given by considering infinitesimal variations of Ω and making use of the Kodaira identity Eq. (2.29) as

$$\Omega\left(z + \delta z\right) = \Omega(z) + \delta z^K\left(\partial_{z^K}\Omega\right)(z) = \Omega(z) - \delta z^K\left(\partial_{z^K}K^{cs}\Omega - \chi^K\right)(z). \tag{B.46}$$

We require that both $\Omega(z)$ and $\Omega(z+\delta z)$ comply with the orientifold constraints Eq. (B.25). Also notice that the complex conjugate to the holomorphic three-form is given by

$$\bar{\Omega}(\bar{z}+\delta\bar{z}) = \bar{\Omega}(\bar{z}) - \delta\bar{z}^K\left(\partial_{\bar{z}^K}\bar{K}^{cs}\bar{\Omega} - \bar{\chi}^K\right)(\bar{z}) = \bar{\Omega}(\bar{z}) - \delta\bar{z}^K\left(\partial_{\bar{z}^K}K^{cs}\bar{\Omega} - \bar{\chi}^K\right)(\bar{z}), \tag{B.47}$$

since $K^{cs} := -i\ln\left(i\int\Omega\wedge\bar{\Omega}\right) = \bar{K}^{cs}$ by the odd properties of the external product on odd-forms. This implies that locally

$$\delta z^K\partial_{z^K}K^{cs} = \delta\bar{z}^K\partial_{\bar{z}^K}K^{cs}, \qquad \delta z^K\sigma^*\chi^K = e^{2i\theta}\delta\bar{z}^K\bar{\chi}^K, \tag{B.48}$$

where we have used that the involution acts trivially on external pieces so that

$$\sigma^*\left[\delta z^K\left(\partial_{z^K}K^{cs}\right)\Omega\right] = \delta z^K\left(\partial_{z^K}K^{cs}\right)\left(\sigma^*\Omega\right),$$
$$\sigma^*\left(\delta z^K\chi^K\right) = \delta z^K\left(\sigma^*\chi^K\right),$$

From the first equality of Eq. (B.48) and recalling that K^{cs} is a Kähler potential, *i.e.* $\partial_{z^K}K^{cs} \neq 0$, we find that for each δz^K either its real or its imaginary part has to be zero. The number of degrees of freedom has been halved and one may choose any basis $z^K = \left(z^k, z^\lambda\right)$ with $\text{Im}z^k = \text{Re}z^\lambda = 0$ for any splitting of $h^{2,1}+1$ in k-components and λ-components. We will keep this splitting generic for now.

These coordinates $z^K = \left(z^k, z^\lambda\right)$ can now be mapped to adapted coordinates of $\mathcal{M}^{cs}_{\mathbb{R}}$ by the embedding [3]

$$\begin{array}{cccc} \rho: & \mathcal{M}^{cs}_{\mathbb{R}} & \hookrightarrow & \mathcal{M}^{cs} \\ & q^K = (z^{\hat{k}}, z^\lambda) & \mapsto & z^K = \left(z^{\hat{k}}, iz^\lambda\right) \end{array}$$

and its metric is given by

$$\rho^*\left(G_{K\bar{L}}dz^K\wedge\star d\bar{z}^L\right) = G_{KL}(q)dq^K\wedge\star dq^L, \qquad \rho^*\left(iG_{K\bar{L}}dz^K\wedge d\bar{z}^L\right) = 0. \tag{B.49}$$

with the period matrices given by restricting

$$\mathcal{M}_{\hat{K}\hat{L}} := \bar{\mathcal{F}}_{\hat{K}\hat{L}} + 2i\frac{(\text{Im}\,\mathcal{F})_{\hat{K}\hat{M}}\,\mathcal{Z}^{\hat{M}}\,(\text{Im}\,\mathcal{F})_{\hat{L}\hat{N}}\,\mathcal{Z}^{\hat{N}}}{\mathcal{Z}^{\hat{M}}\,(\text{Im}\,\mathcal{F})_{\hat{M}\hat{N}}\,\mathcal{Z}^{\hat{N}}}, \tag{B.50}$$

to $\tilde{\mathcal{M}}^{cs}_{\mathbb{R}}$ and applying the constraints

$$\text{Im}(C\mathcal{Z}^{\hat{k}}) = \text{Re}\left(C\mathcal{F}_{\hat{k}}\right) = 0, \quad \text{Re}(C\mathcal{Z}^{\lambda}) = \text{Im}\left(C\mathcal{F}_{\lambda}\right) = 0, \tag{B.51}$$

we find

$$\text{Re}\mathcal{M}_{\sigma\lambda}(q) = \text{Re}\mathcal{M}_{\hat{k}\hat{l}}(q) = \text{Im}\mathcal{M}_{\hat{k}\sigma}(q) = 0, \tag{B.52}$$

Finally, the Kähler potential for the quaternionic manifold is [3, 6]

$$K^{Q} = -2\ln\left[2\int_Y \mathrm{Re}\,(C\Omega)\wedge\star\mathrm{Re}\,(C\Omega)\right]$$
$$= -2\ln 2\left[\mathrm{Re}(CF_\lambda)\mathrm{Im}(CZ^\lambda) - \mathrm{Im}(CF_{\hat{k}})\mathrm{Re}(CZ^{\hat{k}})\right]$$
$$= -\ln e^{-4\phi}. \tag{B.53}$$

For future reference, when writing the supergravity action in term of chiral supermultiplets, it will be convenient to define the coordinates of the quaternionic manifold $\mathcal{M}^{Q}$ in terms of half of the periods of Ω_c [3]

$$N^{\hat{k}} := \int_{\mathcal{X}} \Omega_c \wedge \beta^{\hat{k}} = \xi^{\hat{k}} + 2i\mathrm{Re}(CZ^{\hat{k}}), \tag{B.54}$$

$$U_\lambda := \int_{\mathcal{X}} \Omega_c \wedge \alpha_\lambda = \tilde{\xi}_\lambda + 2i\mathrm{Re}(C\mathcal{F}_\lambda). \tag{B.55}$$

B.3 The Type IIA $\mathcal{N} = 1$ Effective Action

The orientifolded effective action is given by [3]

$$\mathcal{S}^{(4)}_{IIA,\,\mathcal{N}=1} = \int -\frac{1}{2}R\star\mathbf{1} + \frac{1}{2}\mathrm{Im}\mathcal{N}_{\alpha\beta}F^\alpha\wedge\star F^\beta + \frac{1}{2}\mathrm{Re}\mathcal{N}_{\alpha\beta}F^\alpha\wedge F^\beta$$
$$- G_{ab}\,dT^a\wedge\star d\bar{T}^b - h_{uv}dq^u\wedge\star dq^v - V\star\mathbf{1}, \tag{B.56}$$

with the same Freund-Rubin potential

$$V = \frac{e^{4\phi}}{2\mathcal{V}}e_0^2, \tag{B.57}$$

and where

$$h_{uv}d\tilde{q}^u\wedge\star d\tilde{q}^v = d\phi\wedge\star d\phi + G_{KL}(q)dz^K\wedge\star dz^L - \frac{e^{2\phi}}{2}(\mathrm{Im}\mathcal{M})_{\hat{k}\hat{l}}\,d\xi^{\hat{k}}\wedge\star d\xi^{\hat{l}}$$
$$-\frac{e^{2\phi}}{2}(\mathrm{Im}\mathcal{M})^{-1\,\sigma\lambda}\left(d\tilde{\xi}_\sigma - \mathrm{Re}\mathcal{M}_{\hat{k}\sigma}d\xi^{\hat{k}}\right)\wedge\star\left(d\tilde{\xi}_\lambda - \mathrm{Re}\mathcal{M}_{\hat{k}\lambda}d\xi^{\hat{k}}\right), \tag{B.58}$$

is now the metric adapted to the orientifolded quaternionic manifold, of which $\mathcal{M}^{cs}_{\mathbb{R}} \in \mathcal{M}^{Q}$.

Analogously to the type IIB case, we are now presented with the problem of defining appropriate coordinates for the quaternionic manifold. This can be done in a couple different ways. In [3], the author defines coordinates based on the holomorphic periods of the truncated theory. Another approach is to think of the appropriate coordinates as those dual[5] to the Kähler sector of the mirror of $\mathcal{X}$, like in [6]. Below we present the results

- Axio-dilaton: $N^0 := S = s + i\sigma = \xi^0 + 2i\text{Re}(CZ^0) = \int \Omega_c \wedge \beta^0$.
- Kähler moduli: $T^a = v^a + ib^a$, with $a = 1, \dots, h_-^{1,1}$
- Even complex structure moduli: $U_\lambda := u_\lambda + iv_\lambda = \tilde{\xi}_\lambda + 2i\text{Re}(C\mathcal{F}_\lambda) = \int \Omega_c \wedge \alpha_\lambda$.
- Odd complex structure moduli: $N^k := n^k + i\eta^k := \xi^k + 2i\text{Re}(CZ^k) = \int \Omega_c \wedge \beta^k$.

The odd complex structure sector is the dual to the odd Kähler sector in type IIB. To make contact with the same type of theories, we finally make use of the symplectic freedom to choose $N^k = 0$. This implies that $H^3(\mathcal{X}) = H_+^{2,1} \oplus H_+^{3,0}$, and $\lambda = 1, \dots, h_+^{2,1}$. We also remark that the Kähler moduli have been defined in terms of the two-cycle volume moduli v^a in contrast with type IIB case, which was defined in terms of the four-cycle volume moduli τ_α Eq. (2.60).

With this splitting, the Kähler potential of the truncated theory is given by

$$K = K^{ks} + K^Q, \tag{B.59}$$

$$K^{ks} = -\ln\left[\frac{1}{6}\mathcal{K}_{abc}\left(T+\bar{T}\right)^a\left(T+\bar{T}\right)^b\left(T+\bar{T}\right)^c\right], \tag{B.60}$$

$$K^Q = -2\ln\left[2\int_{\mathcal{X}} \text{Re}(C\Omega)\wedge\star\text{Re}(C\Omega)\right], \tag{B.61}$$

and the moduli space is, once again, given locally by the direct product[6]

$$\mathcal{M}^{ks}_{h_-^{1,1}} \times \mathcal{M}^{Q}_{h_+^{2,1}+1}. \tag{B.62}$$

Finally, we would like to put this in the $\mathcal{N} = 1$ gauged supergravity form of [7]

$$\begin{aligned} S^{(4)}_{IIA,\mathcal{N}=1} = -\int & \frac{1}{2}R\star\mathbf{1} + K_{I\bar{J}}DM^I\wedge\star D\bar{M}^{\bar{J}} \\ & + \frac{1}{4}\text{Re}\,\mathcal{N}_{\alpha\beta}F^\alpha\wedge F^\beta + \frac{1}{4}\text{Im}\,\mathcal{N}_{\alpha\beta}F^\alpha\wedge\star F^\beta + V\star\mathbf{1}, \end{aligned} \tag{B.63}$$

---

[5] There are some subtleties due to the broken symplectic invariance in an orientifolded type IIA theory and the choice of orientifolds planes in the type IIB dual, see [3].

[6] Notice that the moduli space will always be given by a direct product of the Kähler sector and the quaternionic sector, no matter the choice of symplectic gauge.

where the scalar potential is given by

$$V = e^{K}\left(K^{I\bar{J}}D_I W D_{\bar{J}}\bar{W} - 3\,|W|^2\right), \tag{B.64}$$

and we have followed the common notation for the chiral fields and Kähler derivatives described below Eq. (2.65). In this case, the scalar potential is non-zero but only contains a cosmological constant-like contribution from the Freund-Rubin flux e_0 Eq. (B.22). Like in the type IIB case, we would like to introduce background fluxes that will allow us to stabilise some moduli.

B.4 Type IIA Flux Compactifications

In this section we will provide the effective action in the case where NSNS and RR background fluxes are turned on. Furthermore, we will consider the case of massive type IIA supergravity by allowing for a non-zero Romans mass parameter m [8]. The massive type IIA ten-dimensional action is

$$\begin{aligned}\mathcal{S}_{IIA-M} = &-\int\left(\frac{1}{2}\hat{R}\star\mathbf{1} + \frac{1}{4}d\hat{\phi}\wedge\star d\hat{\phi} + \frac{1}{4}e^{-\hat{\phi}}\hat{H}_3\wedge\star\hat{H}_3\right)\\ &-\frac{1}{2}\int\left(e^{3\hat{\phi}/2}\hat{F}_2\wedge\star\hat{F}_2 + e^{\hat{\phi}/2}\hat{F}_4\wedge\star\hat{F}_4 + e^{5\hat{\phi}/2}m^2\star\mathbf{1} + \mathcal{L}_{\text{top}}\right)\end{aligned} \tag{B.65}$$

where the topological term now has Romans mass dependent pieces

$$\begin{aligned}\mathcal{L}_{\text{top}} = &\hat{B}_2\wedge d\hat{C}_3\wedge d\hat{C}_3 - (\hat{B}_2)^2\wedge d\hat{C}_3\wedge d\hat{C}_1 + \frac{1}{3}(\hat{B}_2)^3\wedge d\hat{C}_1\wedge d\hat{C}_1\\ &m\left[\frac{1}{3}(\hat{B}_2)^3\wedge d\hat{C}_3 - \frac{1}{4}(\hat{B}_2)^4\wedge d\hat{C}_1 - \frac{1}{3}(\hat{B}_2)^5\right].\end{aligned} \tag{B.66}$$

Similarly, the field strengths are now given by

$$\hat{H}_3 = d\hat{B}_2,\quad \hat{F}_2 = d\hat{C}_1 + m\hat{B}_2,\quad \hat{F}_4 = d\hat{C}_3 - \hat{C}_1\wedge\hat{H}_3 - \frac{m}{2}(\hat{B}_2)^2, \tag{B.67}$$

and the orientifold invariant fluxes will be given by

$$H_3^{(F)} = h_0\beta^0 - h^\lambda\alpha_\lambda,\quad F_2^{(F)} = q^a\omega_a,\quad F_4^F = e_a\tilde{\omega}^a. \tag{B.68}$$

In [7], it was shown that the scalar potential found from dimensional reduction can be obtained in terms of a superpotential

$$W = W_K + W_Q = \int_{\mathcal{X}} e^{-iJ_c} \wedge F^{(F)}_{RR} + \Omega_c \wedge H^{(F)}_3$$
$$= e_0 - ie_a T^a + \frac{1}{2}\mathcal{K}_{abc} q^a T^b T^c - \frac{im}{6}\mathcal{K}_{abc} T^a T^b T^c + S h_0 + U_\lambda h^\lambda, \tag{B.69}$$

and the Kähler potential Eq. (B.59). Here $F^{(F)}_{RR}$ is the poly-form containing all RR background fluxes. The fluxed scalar potential is given by [9]

$$V = \frac{R}{s^4}\left[\frac{\rho_0^2}{2\mathcal{V}} + \frac{G^{ab}\tilde{\rho}_a\tilde{\rho}_b}{8\mathcal{V}} + 2\mathcal{V} G_{ab}\rho^a\rho^b + \frac{\mathcal{V}}{2}\rho_m^2 + \frac{1}{2\mathcal{V}}\left(s^2 h_0^2 - \frac{1}{3}\left(u_\lambda h^\lambda\right)^2\right)\right], \tag{B.70}$$

where $R = 16 e^{2K^{cs}} \mathrm{Re}(Z^0)^4$, and we define the combination $\varphi := e_0 - \xi^0 h_0 + \tilde{\xi}_\lambda h^\lambda$ with

$$\rho_0 = \varphi + e_a b^a + \frac{1}{2}\mathcal{K}_{abc} q^a b^b b^c - \frac{m}{6}\mathcal{K}_{abc} b^a b^b b^c,$$
$$-\rho_a = e_a + \mathcal{K}_{abc} q^b b^c - \frac{m}{2}\mathcal{K}_{abc} b^b b^c, \quad \tilde{\rho}^a = q^a - m b^a, \quad \rho_m = m, \tag{B.71}$$

the dual scalars in the democratic prescription of type IIA [10] and use in Chap. 4.

In contrast with the type IIA case, all moduli appear in the Kähler potential and superpotential. Thus, one might be hopeful that all fields could be stabilised at tree-level. However, on duality grounds we would expect some obstruction to this goal since the type IIB theory required access to non-tree-level corrections to stabilise the Kähler sector. We will study the stabilisation of the type IIA theory in Sect. B.5.

B.5 Moduli Stabilisation in Type IIA Flux Supergravity

In this section we follow the classic analysis of [11]. We will concentrate in studying the stabilisation of supersymmetric vacua configurations of type IIA. We will find that we are able to stabilise the dilaton, a linear combinations of axions in φ, the complex structure saxions u_λ and the Kähler sector, leaving $h^{2,1}$ complex structure axions flat. To begin, we look for a solution to the F-term equations

$$D_T W = D_S W = D_U W = 0. \tag{B.72}$$

Let us start with the quaternionic sector. The F-term for the axio-dilaton is

$$D_S W = h_0 - \frac{2iW}{e^{\sqrt{K^Q}}}\mathrm{Im}\,(C F_0) = h_0 - 2iW e^{2\phi}\mathrm{Im}\,(C F_0) = 0, \tag{B.73}$$

and for the complex moduli we find

$$D_{U^\lambda} = h^\lambda + \frac{2iW}{e^{\sqrt{K^Q}}}\mathrm{Im}(CZ^\lambda) = h^\lambda + 2iWe^{2\phi}\mathrm{Im}(CZ^\lambda) = 0, \tag{B.74}$$

where we have used the definitions of the chiral superfields in terms of the periods of Ω_c so that

$$\partial_S = \partial_{2\mathrm{Re}(CZ^0)} \frac{\partial 2\mathrm{Re}\left(CZ^0\right)}{\partial S} = -i\partial_{2\mathrm{Re}(CZ^0)}, \tag{B.75}$$

$$\partial_{U_\lambda} = \partial_{2\mathrm{Re}(CF_\lambda)} \frac{\partial 2\mathrm{Re}\left(CF_\lambda\right)}{\partial U^\lambda} = -i\partial_{2\mathrm{Re}(CF_\lambda)}. \tag{B.76}$$

Noting that the dilaton is real-valued, the imaginary parts of the quaternionic F-terms lead to the same condition

$$\mathrm{Re}W = 0 \quad \rightarrow \quad h_0\sigma + \nu_\lambda h^\lambda + \mathrm{Re}W_K = 0 \;\; \text{or} \;\; \varphi = -\mathrm{Re}W_K. \tag{B.77}$$

These identical equations allows us to stabilise the dilaton and φ field at the minimum of the potential once we solve for the Kähler and complex structure fields. The degeneracy in these F-terms can be seen from the fact that we have $h^{2\,1} + 1$ complex equations and only $h^{2,1} + 1$ real fluxes in $\left(h_0, h^\lambda\right)$, so there is not info degrees of freedom to stabilise the whole system.

Furthermore, we note that, after solving the previous F-terms

$$W = \mathrm{Re}W + i\mathrm{Im}W = i\mathrm{Im}W, \tag{B.78}$$

and thus $\mathrm{Im}W = 0$ implies a vanishing superpotential, which is inconsistent with the presence of any fluxes. This implies that the only solution to the real part of Eq. (B.73) is given by

$$h_0 + 2e^{2\phi}\mathrm{Im}W\,\mathrm{Im}(CF_0) = 0 \quad \rightarrow \quad e^{-\phi} = -2\frac{\mathrm{Im}W}{Q_0}, \tag{B.79}$$

with $Q_0 := e^{-\sqrt{K^{cs}}}\frac{h_0}{\mathrm{Im}(F_0)}$. Similarly, for the complex structure sector we find λ equations

$$h^\lambda - 2e^{2\phi}\mathrm{Im}W\,\mathrm{Im}(CZ^\lambda) = 0 \quad \rightarrow \quad \frac{e^{-\phi}}{2\mathrm{Im}W} = e^{-\sqrt{K^{cs}}}\frac{h^\lambda}{\mathrm{Im}(Z^\lambda)} := -Q_0. \tag{B.80}$$

Then the chain of $h^{2,1}$ equations

$$e^{-\sqrt{K^{cs}}}\frac{h^1}{\mathrm{Im}(Z^1)} = e^{-\sqrt{K^{cs}}}\frac{h^2}{\mathrm{Im}(Z^2)} = \ldots = e^{-\sqrt{K^{cs}}}\frac{h^{h^{2,1}}}{\mathrm{Im}(Z^{h^{2,1}})} := -Q_0, \tag{B.81}$$

fixes the complex structure sector. On top of these, Eq. (B.80) will stabilise the ten-dimensional dilaton $\hat{\phi}$ after the Kähler sector is fixed, given that the four-dimensional dilaton ϕ contains a volume factor in its definition.

Before turning to the Kähler moduli, let us note that when the previous equations are satisfied the superpotential can be written in terms of its Kähler moduli only as

$$W(T^a, S, U^k) = -i\mathrm{Im}W_K\left(T^a\right), \tag{B.82}$$

meaning the Kähler sector completely decouples from the complex structure sector and axio-dilaton. We can see this by multiplying Eq. (B.73) by $\mathrm{Re}(CZ^0)$ and Eq. (B.74) by $\mathrm{Re}(CF_\lambda)$ and adding them to obtain

$$\begin{aligned} &-\frac{1}{2}\left(h_0\sigma + h^\lambda v_\lambda\right) - iWe^{2\phi}\left[2\left(\mathrm{Re}(CF_\lambda)\mathrm{Im}(CZ^\lambda) - \mathrm{Im}(CF_{\hat{k}})\mathrm{Re}(CZ^{\hat{k}})\right)\right] \\ &= -\frac{1}{2}\mathrm{Im}W^Q - iW = 0. \end{aligned} \tag{B.83}$$

The F-terms for the Kähler moduli are then

$$\begin{aligned} D_{T^a}W &= -ie_a + \mathcal{K}_{abc}q^bT^c - \frac{im}{2}\mathcal{K}_{abc}T^bT^c + i\frac{\mathrm{Im}W_K}{2\mathcal{V}}\mathcal{K}_{abc}\left(T+\bar{T}\right)^b\left(T+\bar{T}\right)^c \\ &= -ie_a + \mathcal{K}_{abc}q^b(v-ib)^c - \frac{im}{2}\mathcal{K}_{abc}(v^bv^c - b^bb^c - 2ib^bv^c) \\ &\quad + i\frac{2\mathrm{Im}W_K}{\mathcal{V}}\mathcal{K}_{abc}v^bv^c = 0. \end{aligned} \tag{B.84}$$

The real part leads to

$$\mathcal{K}_{abc}v^c\left(q^b - mb^b\right) = 0. \tag{B.85}$$

Since $\mathcal{K}_{abc}v^c \propto G_{ab}$, the Kähler metric, for any given row there must at least be a non-vanishing component so that the metric is invertible. Thus, the object inside the parentheses must vanish yielding the stabilisation for the Kähler axions

$$b^a = \frac{q^a}{m}. \tag{B.86}$$

The imaginary part of the F-term for the Kähler sector gives

$$\begin{aligned} &\mathcal{K}_{def}v^dv^ev^f\left(4me_a + 2\mathcal{K}_{abc}q^bq^c + 3m^2\mathcal{K}_{abc}v^bv^c\right) + \\ &\mathcal{K}_{abc}v^bv^c\left(6me_dv^d + 3\mathcal{K}_{def}q^dq^ev^f\right) = 0, \end{aligned} \tag{B.87}$$

this gives $h^{1,1}$ real-equations that are quadratic in v^a and allow us to stabilise them. We can also rewrite this equation in terms of the dual scalars from Eq. (B.71), after tracing over v^a, as

$$v^a \left(10me_a + 5\mathcal{K}_{abc}q^b q^c + 3m^2\mathcal{K}_{abc}v^b v^c\right) = v^a \left(10\rho_a + 3m^2\mathcal{K}_{abc}v^b v^c\right) = 0, \tag{B.88}$$

Thus, the Kähler sector is fully stabilised, as opposed to the complex structure sector where only the saxions are fixed at the minimum. One can also understand this as the Kähler sector having $2 \cdot h^{1,1}_-$ fluxes, (e_a, q^a), compared to the $h^{2,1}_+$ fluxes of the complex structure sector, h^λ.

In contrast to the type IIB moduli stabilisation on Sect. 2.4, we have been able to stabilise the Kähler sector, the axio-dilaton and the saxions of the complex structure sector. By duality arguments, however, we see that the complex structure saxions map to 4-volume cycles in the dual type IIB [3, 5], which had flat profiles at the same level of the perturbative expansion. In [12], it was shown that turning on any of the h^λ fluxes required to stabilise the complex structure sector leads to a half-flat manifold and to non-geometric fluxes in the dual type IIB setting. This fluxes are not fully understood and present a problem regarding the trustability of solutions that employ them as means to stabilise moduli.

In Sect. 4.5, we present a no-go theorem for the embedding of clockwork into perturbative type IIA theories. We use a form of the potential that formally contains this fluxes, but the argument is entirely focused on the Kähler sector and the impossibility of generating the clockwork mechanism through those. The argument only requires a decoupling of the Kähler and complex structure sector, which is true at tree-level in any case, to go through. Therefore, the trustability issues are not a factor for our discussion.

References

1. Polchinski J (2007) String theory. vol 2: superstring theory and beyond. Cambridge Monographs on Mathematical Physics, Cambridge University Press. https://doi.org/10.1017/CBO978051161812310.1017/CBO9780511618123
2. Louis J, Micu A (2002) Type 2 theories compactified on Calabi-Yau threefolds in the presence of background fluxes. Nucl Phys B 635:395. https://doi.org/10.1016/S0550-3213(02)00338-3, https://arxiv.org/abs/hep-th/0202168
3. Grimm TW (2005) The effective action of Type II Calabi-Yau orientifolds. Fortsch Phys 53:1179. https://doi.org/10.1002/prop.200510253, https://arxiv.org/abs/hep-th/0507153
4. Dabholkar A (1997) Lectures on orientifolds and duality. In: ICTP summer school in high-energy physics and cosmology, vol 6, pp 128–191. https://arxiv.org/abs/hep-th/9804208
5. Hosono S, Klemm A, Theisen S (1994) Lectures on mirror symmetry. Lect Notes Phys 436:235. https://doi.org/10.1007/3-540-58453-6_13, https://arxiv.org/abs/hep-th/9403096
6. Palti E, Tasinato G, Ward J (2008) WEAKLY-coupled IIA Flux Compactifications. JHEP06:084. https://doi.org/10.1088/1126-6708/2008/06/084, https://arxiv.org/abs/0804.1248
7. Grimm TW, Louis J (2005) The effective action of Type IIA Calabi-Yau orientifolds. Nucl Phys B 718:153. https://doi.org/10.1016/j.nuclphysb.2005.04.007, https://arxiv.org/abs/hep-th/0412277
8. Romans LJ (1986) Massive N = 2a supergravity in ten-dimensions. Phys Lett B 169:374. https://doi.org/10.1016/0370-2693(86)90375-8

9. Bielleman S, Ibanez LE, Valenzuela I. Minkowski 3-forms, flux string vacua, axion stability and naturalness. JHEP 12:119. https://doi.org/10.1007/JHEP12(2015)119, https://arxiv.org/abs/1507.06793
10. Carta F, Marchesano F, Staessens W, Zoccarato G (2016) Open string multi-branched and Kähler potentials. JHEP 09:062. https://doi.org/10.1007/JHEP09(2016)062, https://arxiv.org/abs/1606.00508
11. DeWolfe O, Giryavets A, Kachru S, Taylor W (2005) Type IIA moduli stabilization. JHEP 07:066. https://doi.org/10.1088/1126-6708/2005/07/066, https://arxiv.org/abs/hep-th/0505160
12. Acharya BS, Benini F, Valandro R (2007) Fixing moduli in exact Type IIA flux vacua. JHEP 02:018. https://doi.org/10.1088/1126-6708/2007/02/018, https://arxiv.org/abs/hep-th/0607223